Diagnostics and Therapeutic Advances in GI Malignancies

This series will highlight the recent innovations in the diagnostics and therapeutic strategies for different Gastrointestinal (GI) cancers.

Gastrointestinal cancers are a group of cancers that affect the digestive system and include gastric cancer, colorectal cancer, liver cancer, esophageal cancer, and pancreatic cancer. GI cancers are the leading health problem in the world and their burden is increasing in many countries. This heavy burden is due to the lack of effective early detection methods and to the emergence of chemoradioresistance. Attempts at improving the outcome of GI cancers by incorporating cytotoxic agents such as chemo drugs have been so far disappointing. These results indicate that the main challenge remains in the primary resistance of GI cancer cells to chemotherapy in the majority of patients. Therefore, improvement in the outcomes of these malignancies is dependent on the introduction of new agents that can modulate the intrinsic and acquired mechanisms of resistance.

The increased understanding of the biology, metabolism, genetic, epigenetic, and molecular pathways dysregulated in GI cancers has revealed the complexity of the mechanisms implicated in tumor development. These include alterations in the expression of key oncogenic or tumor suppressive miRNAs, modifications in methylation patterns, the upregulation of key oncogenic kinases, etc.

The individual books in this series will focus on the genetic basis of each gastrointestinal cancers, molecular pathophysiology, and different biomarkers to estimate cancer risk, detection of cancer at microscopic dimensions, and suitable and effectiveness of the therapies. In addition, the volumes will discuss the role of various signaling molecules/pathways and transcriptional factors in the regulation of the tumor microenvironment and effect on the tumor growth.

Lastly, it will elaborate the use of molecularly targeted drugs that have been proven to be effective for the treatment of GI cancers, with a focus on the emerging strategies.

This edition will provide researchers and physicians with novel ideas and perspectives for future research that translates the bench to the bedside.

More information about this series at http://www.springer.com/series/16343

Pallaval Veera Bramhachari •
Nageswara Rao Reddy Neelapu
Editors

Recent Advancements in Biomarkers and Early Detection of Gastrointestinal Cancers

Springer

Editors
Pallaval Veera Bramhachari
Department of Biotechnology
Krishna University
Machilipatnam, Andhra Pradesh, India

Nageswara Rao Reddy Neelapu
Department of Biochemistry and Bioinformatics
GITAM Institute of Science, Gandhi Institute of Technology and Management (GITAM)
(Deemed-to-be-University)
Visakhapatnam, Andhra Pradesh, India

ISSN 2662-2688 ISSN 2662-2696 (electronic)
Diagnostics and Therapeutic Advances in GI Malignancies
ISBN 978-981-15-4433-0 ISBN 978-981-15-4431-6 (eBook)
https://doi.org/10.1007/978-981-15-4431-6

This Springer imprint is published by the registered company Springer Nature Singapore Pte Ltd.
The registered company address is: 152 Beach Road, #21-01/04 Gateway East, Singapore 189721, Singapore

Preface

In cancer, the novel biomarkers include substances produced by the cancer tissue or by other cells reacting to cancer in the body. They can be predominantly helpful in detecting and diagnosing, predicting responses to therapy, tracking treatment results or cancer growth, and finally determining whether a cancer has returned subsequent remission. There are three main tests: (1) Genetic tests search for abnormal changes and mutations, together with extra, missing, or erroneously placed genes. (2) Biochemical tests determine if there are too many proteins or if proteins are overactive. (3) Karyotyping identifies abnormal changes within chromosomes.

Gastrointestinal cancers (GI) signify an essential portion of global public health concern with millions of deaths annually. GI cancers are biologically and genetically heterogeneous, with a poorly understood carcinogenesis at the molecular level. Although cancer incidence is declining, the outcome of patients with GI cancers remains hugely dismal. Thus, detection at an early stage utilizing functional screening approaches, selection of an appropriate treatment plan, and effective monitoring is fundamental to reduce GI cancer mortalities. Nonetheless, there have been tremendous advancements in the multidisciplinary management of GI cancers in the past decade. The growing number of new and improved targeted agents and efficacious combination regimens yielding substantial clinical benefits at both early and late stages of the disease proved this. Moreover, breakthroughs in molecular profiling, cancer immunology, early stage detection, and novel diagnostic techniques have led to accelerated strides in GI cancer research. As the field of GI malignancies is continuously evolving, community oncologists must strive hard to keep abreast of the latest developments in novel biomarker research and resolve new issues in optimizing the management and detection of GI cancers.

In the field of cancer biology, the researchers continue to make noteworthy and exhilarating contributions to understand the fundamental biology of the GI cancers. Yet, the practical translational applications of this fascinating and enthralling area of science is little disappointing with regard to the recurrent viral outbreaks. These events underscore the need for concerted efforts to develop and implement new interventions while continuing to invest in proven public health measures.

Considering the high mortality rate, tremendous effort has been directed to address the urgent need for the discovery of effective early diagnostic tools, efficient therapeutic targets, and treatment monitoring markers for GI malignancies.

Biomarkers are one of the favorite tools with several potential applications in various aspects of clinical management of cancers. The biomarkers in GI cancer research primarily focus on the patient's unique clinical characteristics. Additionally, they facilitated the researchers and healthcare professionals to better support GI cancer patients through (1) understanding how to prevent different diseases, (2) diagnosing the sternness or stage of an illness, (3) helping to inform a patient with treatment options, and (4) determining the probability if the disease returns. However, the identification of biomarkers and continuing discovery of new ones mark the clear evolution of how clinicians and patients can effectively determine personalized treatments. Therefore, identification of novel biomarkers on the basis of clinical information (serology, metabolic and biochemical data) is mandatory, and furthermore comprehensive genome analysis could undeniably improve the diagnosis, prognosis, prediction of recurrence, and treatment response for GI cancers.

A plethora of biomarkers have been previously studied in GI cancers, of which only a handful have found their way from bench to bed. Nonetheless, there is a growing list of emerging markers with promising clinical results that need to be validated for routine clinical applications, and current data are insufficient to recommend them as part of the clinical guidelines. Biomarkers must be rigorously tested and validated in clinical studies. Biomarker research also needs to be interpreted carefully, so that the patients are not excluded from receiving potentially helpful medicines. Despite of various challenges associated with the discovery of novel biomarkers and testing for clinical studies, biomarkers are deemed to be critical components of cancer research.

The scientific advances are producing potential new biomarkers for the early detection of cancer and improved disease management. Advances in genomics, proteomics, and molecular pathology have produced many candidate biomarkers with the potential to impact clinical care for GI cancers. These novel biomarkers particularly emphasizes the early detection of cancer, prediction of disease outcomes, and how the methods for such discovery are being used in personalized medicine. Understanding the basic biology of cancer and identifying new biomarkers could be critical for the growth and progression of particular GI cancers. Also finding new biomarkers for GI cancers could be of great value in the effort to determine which pathways are important to target the new investigational research.

We affirm that this book would provide enough insights into the current understanding of the prognosis and prediction of biomarkers by profiling microRNA, circulating microRNAs, serum microRNA, and plasma microRNA for diagnosing early onset of gastric cancer. This book is an attempt to compile the novel information available on recent advancements on the breakthrough technologies such as ultra-sensitive nanochip, nanosensors, nanodevices, biosensors, electrochemical biosensors, optical biosensors, and DNA biosensors for the early diagnosis of gastrointestinal cancer. The book also elucidates a comprehensive yet a representative description of a large number of challenges associated with the discoveries and the role of molecular and biochemical biomarkers akin to volatile biomarkers, serum biomarkers, predictive and prognostic molecular markers for the early detection of gastrointestinal cancers. This book could be an essential reading for the novice and experts in the field of cancer biology, cancer immunodiagnostics, including latest developments in biomarker research. With these objectives in mind, the content of this textbook has been arranged in a logical progression from fundamental to more advanced concepts. Finally, this book also outlines the most advanced biomarker techniques used in diagnostics of GI cancers and also primarily focuses on advancements of biomarker development research and management. Development of these biomarkers in the field of cancer treatment is expected to greatly contribute to the progress of cancer, selection of appropriate therapeutic strategies, and efficient follow-up programs.

We hope that this book stimulates your creativity and wishes you success in your experiments. This book is a stunning reflection of the seriousness with which the several scientific minds are dedicated to the welfare of the scientific community. We are extremely thankful to the contributors for paying continuous attention to our request and showing faith in our capabilities. We shall always remain highly obliged to all of them forever. These words cannot justify the worthiness of their efforts. We successfully compiled our creative and thoughtful research work due to genuine concern and painstaking effort of many more well-wishers whose names are not mentioned, but they are still in our heart. So, the reward is surely worth for their efforts. We and the contributing authors hope from the bottom of our hearts that this book will be a good guidebook and compass for research studies on novel biomarkers for the early detection of gastrointestinal cancers.

Machilipatnam, Andhra Pradesh, India — Pallaval Veera Bramhachari
Visakhapatnam, Andhra Pradesh, India — Nageswara Rao Reddy Neelapu

Acknowledgements

Our sincere thanks are extended to all the academicians and scientists who have contributed chapters and happily agreed to share their work on *Recent Advancements in Biomarkers and Early Detection of Gastrointestinal Cancers* in this volume.

This book is a stunning reflection of the seriousness with which the several scientific minds are dedicated from the scientific community. We are extremely thankful to the contributors for paying continuous attention to our request and showing faith in my competencies and capabilities. We shall always remain highly obliged to all of them forever. These words cannot justify the worthiness of their efforts. We appreciate the excellent work of the authors and co-authors who were invited to contribute chapters in this book. The credit for making this book a reality goes to them. As editors and our review team for the chapters especially appreciate sharing expertise with the contributors. Each chapter is informative and written as a stand-alone, so the reader can begin anywhere in the book depending upon his/her interests and needs.

At the same time, we also express our deepest gratitude to our family members for their kind support which has prompted us to complete the assignment on time. We are also thankful to the Department of Biotechnology, Krishna University and the Department of Biochemistry and Bioinformatics, GITAM (Deemed-to-be-University), AP, India, for the support. We are equally thankful to the Springer Nature Publishing group for their full cooperation during the peer review and production of the volume.

We are thankful to our beloved teachers and mentors, for their constant support and motivations at all stages of progress.

About the Book

This book illustrates the importance or need for the early detection or diagnosis of gastrointestinal cancer. This book *Recent Advancements in Biomarkers and Early Detection of Gastrointestinal Cancers* provides information on discovery of biomarkers by profiling microRNA, circulating microRNAs, serum microRNA, and plasma microRNA for diagnosing early onset of gastric cancer. Further, it provides breakthrough technologies such as ultra-sensitive nanochip, nanosensors, nanodevices, biosensors, electrochemical biosensors, optical biosensors, and DNA biosensors for the early diagnosis of gastrointestinal cancer. It also describes the discoveries and the role of molecular markers or biomarkers like volatile biomarkers, serum biomarkers, predictive and prognostic molecular markers for the early detection of gastrointestinal cancer. GWAS, big data analytics, computation biology, and systems biology approaches can be used to discover and develop diagnostics and therapeutics for gastrointestinal cancer. In closing, the book provides comprehensive information, inspiration, and advanced clinical applications on early diagnosis and detection of gastrointestinal cancers aiming towards personalized medicine to treat cancer.

Contents

Editors and Contributors

About the Editors

Pallaval Veera Bramhachari is currently a Senior Faculty member of the Department of Biotechnology at Krishna University, Machilipatnam, India. Earlier, I served as Postdoctoral Research Scientist at the Bacterial Pathogenesis Laboratory, Queensland Institute of Medical Research (QIMR-Berghofer Research Center), Brisbane, Australia, and DBT Postdoctoral Fellow at the Department of Microbiology and Cell Biology, Indian Institute of Science (IISc), Bangalore. I completed my Ph.D. in Microbiology from Goa University, India. My major research interests are the investigation on the applied research in microbiology and cell biology including biochemical mechanism of bacterial-biofilm formations, structure–function relationship of bacterial rhamnolipids, and basic and translational cancer research. I published more than 125 research articles including 5 international books in prestigious peer-reviewed international Scopus-indexed journals and presented 50 abstracts at various national and international conferences. I served as the Editorial Board member and as Reviewer for a number of national and international journals. I am a member of many international scientific societies and organizations, most importantly, Indian Science Congress and Society of Biological Chemists, India. I successfully completed two major research projects. I was awarded Travel Scholarship from QIMR, Australia, in 2007 for attending 4th Indo-Australian Biotechnology Conference at Brisbane, Australia, and awarded with Young Scientist

Travel Fellowship (2007) from the DST, Govt. of India, for attending XVII Lancefield International Symposium at Porto Heli, Greece (2008). Dr. Bramhachari was conferred with various prestigious awards, notably, Science Education Research Board (SERB), Government of India-Young Scientist award (2011) with a Research Project, Nominated as Associate Fellow of Andhra Pradesh Academy of Sciences (APAS) (2016), MASTER TRAINER: Andhra Pradesh, English Communications Skills Project by (British Council and APSCHE, 2017), Andhra Pradesh State Best Scientist award (APCOST, 2017), Dr. V. Ramalingaswamy Memorial award for Biomedical Sciences (2019) by Science City of Andhra Pradesh. I also obtained two Indian patents in 2017. I have more than 13 years of teaching and research experience at the university level.

Nageswara Rao Reddy Neelapu is currently an Assistant Professor in Bioinformatics Division at the Department of Biochemistry and Bioinformatics, Institute of Science, Gandhi Institute of Technology and Management (GITAM), Deemed-to-be-University, Visakhapatnam, India. He obtained his Masters in Life Science from Pondicherry Central University, Pondicherry, and Ph.D. in Evolutionary Bioinformatics from Andhra University, Visakhapatnam, India. His research interest is in Bioinformatics (with specialization in drug target discovery, HLA-based peptide vaccine design, genome analysis, phylogeny and evolution, structure–function relationship of proteins, and exploring next-generation sequencing data), Fungal Biology, and Cancer. Furthermore, he was a recipient of several fellowships like CSIR-UGC Junior Research Fellowship; and DST-DAAD Visiting Fellowship to Germany. He has published more than 50 research articles in prestigious peer-reviewed international Scopus-indexed journals and presented 18 abstracts at various national and international conferences. He has served on editorial board for a number of national and international journals, most importantly, Artificial Intelligence in Gastroenterology of **Baishideng Publishing Group Inc.** He is a member of many international scientific societies and organizations, most importantly, Indian Science Congress & Society of Biological Chemists, India. He also served as reviewer for a number of international journals. He has more than 12 years of teaching experience at the university level in Computational Biology, Bioinformatics, Computing, Genomics, Cancer Informatics, Immunology and Immunoinformatics, and Molecular Phylogeny. He has guided three doctoral students for their Ph.D. and filed an Indian patent in 2018.

Contributors

Pallaval Veera Bramhachari Department of Biotechnology, Krishna University, Machilipatnam, Andhra Pradesh, India

K. V. Chamundeswaramma Department of Zoology, Government College for Men, Anantapur, Andhra Pradesh, India

Manoj Kumar Gupta Department of Biotechnology and Bioinformatics, Yogi Vemana University, Kadapa, Andhra Pradesh, India
Department of Biochemistry, Yogi Vemana University, Kadapa, Andhra Pradesh, India

Achanta Jagadeesh Tumor Microenvironment Global Core Research Center, College of Pharmacy, Seoul National University, Seoul, South Korea

Kummetha Jagadish GVK BIO, Hyderabad, Telangana State, India

M. Kiran Kumar Department of Biotechnology, College of Science and Technology, Andhra University, Visakhapatnam, Andhra Pradesh, India

Koppula Satish Kumar Department of Biotechnology, Adikavi Nannaya University, Rajamahendravaram, Andhra Pradesh, India

Ravi Chandra Pavan Kumar Sri-Tirumala-Peddiniti Department of Biochemistry and Bioinformatics, GITAM Institute of Science, Gandhi Institute of Technology and Management (GITAM) (Deemed-to-be-University), Visakhapatnam, Andhra Pradesh, India

Sneha Mulpuri Department of Biotechnology, Andhra University, Visakhapatnam, Andhra Pradesh, India

B. Prathap Naidu Department of Biotechnology, Krishna University, Machilipatnam, Andhra Pradesh, India

Deepthi Nammi Department of Biochemistry and Bioinformatics, GITAM Institute of Science, Gandhi Institute of Technology and Management (GITAM) (Deemed-to-be-University), Visakhapatnam, Andhra Pradesh, India

A. M. V. N. Prathyusha Department of Biotechnology, Krishna University, Machilipatnam, Andhra Pradesh, India

Jinka Rajeswari Department of Biochemistry, Acharya Nagarjuna University, Guntur, Andhra Pradesh, India

Nageswara Rao Reddy Neelapu Department of Biochemistry and Bioinformatics, GITAM Institute of Science, Gandhi Institute of Technology and Management (GITAM) (Deemed-to-be-University), Visakhapatnam, Andhra Pradesh, India

Pamuru Ramachandra Reddy Department of Biotechnology and Bioinformatics, Yogi Vemana University, Kadapa, Andhra Pradesh, India
Department of Biochemistry, Yogi Vemana University, Kadapa, Andhra Pradesh, India

Pavani Sanapala Department of Biotechnology, Andhra University, Visakhapatnam, Andhra Pradesh, India

Prakash C. Sharma University School of Biotechnology, Guru Gobind Singh Indraprastha University, New Delhi, India

G. Mohana Sheela Department of Biotechnology, Krishna University, Machilipatnam, AP, India

Pola Sudhakar Department of Biotechnology, Andhra University, Visakhapatnam, Andhra Pradesh, India

Surekha Challa Department of Biochemistry and Bioinformatics, GITAM Institute of Science, Gandhi Institute of Technology and Management (GITAM) (Deemed-to-be-University), Visakhapatnam, Andhra Pradesh, India

Ramakrishna Vadde Department of Biotechnology and Bioinformatics, Yogi Vemana University, Kadapa, Andhra Pradesh, India

Renu Verma University School of Biotechnology, Guru Gobind Singh Indraprastha University, New Delhi, India

Abbreviations

5,6-dihydroxy-5,6-dihydrothymidine	Thymidine glycol
ACRG	Asian Cancer Research Group
ACS	American Cancer Society
ALP	Alkaline Phosphatase
ALT	Alanine Transaminase
AMPK	5′-AMP-activated protein kinase
APC	Adenomatous polyposis coli
ASR	Analyte-specific reagent
AST	Serum Aspartate Transaminase
BCFA	Branched chain fatty acids
BCOC	Be Clear on Cancer
BUN	Blood Urea Nitrogen
CAG	Chronic atrophic gastritis
CC	Creatinine Clearance
CCC	Cytoplasmic cell-adhesion complex
CDSS	Clinical Decision Support System
CDX-2	Caudal type homeobox-2
CE	Chromoendoscopy
CEA	Carcinoembryonic antigen
CE-MS	Capillary electrophoresis-mass spectroscopy
cfDNA	Circulating cell-free DNA
CIP	Cancer Imaging Program
CTCs	Circulaing Blood Cells
CIS	Chromosomal instability
CK	Creatine Kinase
CKs	Cytokeratins
CLE	Confocal laser endomicroscopy
CLIA	Clinical Laboratory Improvement Amendments

CMS	Centers for Medicare and Medicaid Services
CNT	Carbon Nanotubes
CNV	Copy number variation
CPCA	Consensus PCA
CRC	Colorectal cancer
CRP	C-reactive protein
CSB	Cancer systems biology
CSG	Chronic superficial gastritis
CT	Cytoplasmic tail
CT scan	Computerized tomography scan
CTCs	Circulating tumor cells
ctDNA	Circulating tumor DNA
cTn	Cardiac Troponin
CYP2D6	Cytochrome P450-2D6
ddPCR	Droplet digital PCR
DGC	De novo diffuse-type Gastric Cancer
DNMT1	DNA methyltransferases
DNMT3A	DNA methyltransferase-3A
DSS	Decision support system
DYS	Gastric dysplasia
EGC	Early gastric cancer
EGTM	European Group on Tumor Markers
EMT	Epithelial–mesenchymal transition
EpCAM	Epithelial cell adhesion molecule
ESCC	Esophageal squamous cell carcinoma
EZH2	Enhancer of zeste homolog 2
FDG-PET	Fluorodeoxyglucose Positron Emission Tomography
FICE	Flexible spectral imaging color enhancement
FISH	Fluorescent in situ hybridization
FMo3	Monooxygenase-3 or CYP1A2
FMo3	Flavin-Containing Monooxygenase-3 or CYP1A2
FOBT	Fecal occult blood testing
FOXO4	Forkhead box protein O4
FRET	Forster Resonance Energy Transfer Biosensors
FTICR	Fourier transform ion cyclotron resonance
FT-IR	Fourier Transform Infrared
FXR	Farnesoid X receptor
GCA	Graded clustering analysis
GC-MS	Gas chromatography-mass spectroscopy

GFR	Glomerular Filtration Rate
GGT	Gamma-Glutamyl Transferase
GI	Gastrointestinal cancer
GNCA	Gastric cardio adenocarcinoma
GO	Graphene Oxide
GOBIOM	GVK Bio Online Biomarker Database
GSK-3β	Glycogen synthase kinase 3β
GWAS	Genome-wide association studies
HbA1c	Glycosylated hemoglobin
HER2	Human epidermal growth factor receptor 2
HGF	Hepatic growth factor
HNF4A	Hepatocyte nuclear factor 4 alpha
HO*	Hydroxyl radicals
HPLC-MS	High performance mass spectroscopy
HR	Novel high-resolution virtual chromoendoscopy
IGC	Intestinal-type Gastric Cancer
IGCLC	The International Gastric Cancer Linkage Consortium
IHC	Immunohistochemistry
IL1B-31T	Interleukin-1 beta
IM	Intestinal metaplasia
IVD	In vitro diagnostic
IVUS	Intravascular Ultrasonography
JAK2	Janus kinase 2
KIM-1	Kidney Injury Molecule-1
LC-MS	Liquid Chromatography-Mass Spectrometry
LDH	Lactate Dehydrogenase
lncRNAs	Long noncoding RNAs
LOD	Limit of detection
LOH	Loss of heterozygosity
LPA	Lysophosphatidic acid
LysoPC	4-Lysophosphatidylcholines
MA-CD	Magnetic cyclodextrin polymer
MALT	Mucosa-associated lymphoid tissue
MFCs	Microbial fuel cells biosensor
MIAMOD	Mortality and Incidence Analysis Model
MP	Microparticles
MRI	Magnetic resonance imaging
MRS	Magnetic resonance spectroscopy
MS	Mass spectrometry
MSI	Microsatellite instability
MSP	Methylation-specific PCR

MudPIT	Multidimensional Protein Identification Technology
MVDA	Multivariate data analysis
NAG	*N*-acetyl-β-D-glucosaminidase
NBI	Narrow-band imaging
NBIA	National Biomedical Imaging Archive
NGAL	Neutrophil Gelatinase-Associated Lipocalin
NM	Nuclear medicine
OPLS-DA	Orthogonal partial least squares discriminant analysis
OSC	Orthogonal signal correction
PCA	Principal component analysis
PD-L	Programmed death ligand
PET	Positron emission tomography
PFAA	Plasma-free amino acid
PIAMOD	Prevalence and Incidence Analysis Model
PK	Pharmacokinetic measurements
PLS-DA	Partial least squares discriminant analysis
POMA	Pipeline of outliner MicroRNA analysis
PPIs	Protein–protein interactions
prGO	Partially reduced Graphene Oxide
pTNM	Pathologic tumor-node-metastasis
PTPRC or CD45	Protein tyrosine phosphatase
PTPRCAP	Protein tyrosine phosphatase receptor type C-associated protein
PTR-MS	Proton transfer reaction mass spectrometry
QCA	Quantitative Coronary Angiography
qPCR	Quantitative polymerase chain reaction
qRT-PCR	Quantitative reverse transcriptase polymerase chain reaction
QTLs	Quantitative trait loci
q-TOF	Quadrupole time-of-flight
RFA	Recurrent focal amplifications
rGO	Reduced graphene oxide
RH	Rheumatoid arthritis
RhoA	Ras homolog family member A
ROC	Receiver operating characteristics
ROCK1	Rho-associated coiled-coil containing protein kinase 1
RTKs	Receptor tyrosine kinases
SAGE	Serial analysis of gene expression
SC	Serum creatinine
SELDI	Surface-enhanced laser desorption/ionization

SELDI-TOF	Surface-enhanced laser desorption/ionization Time-of-flight
SFKs	Src family kinases
SIFT-MS	Selected ion flow tube mass spectrometry
SNPs	Single nucleotide polymorphisms
SOPs	Standard operating procedures
SPECT	Single Photon Emission Computed Tomography
SPME	Solid-phase microextraction
SQUID	Superconducting quantum interference device
SVM	Support vector machine
TCA	Tricarboxylic acid
TCGA	The Cancer Genome Atlas
TCIA	The Cancer Imaging Archive
TFF1	Trefoil factor 1
TIMP1	Metallopeptidase inhibitor 1
TOF	Time-of-flight mass spectrometry
TSH	Thyroid-stimulating hormone
TTR	Time to results
UPLC-MS	Ultra performance mass spectroscopy
US	Ultrasound
USPSTF	United States Preventive Services Task Force
VEGF	Vascular endothelial growth factor
VOCs	Volatile organic compounds
XRCC1	X-ray repair cross-complementing group 1
ZIKV	Zika virus antigens

Part I
Emerging Biomarkers for Gastrointestinal Cancer

Chapter 1
Potential Role of Biomarkers, Biosensors, Technologies, and Computational Methods in Early Detection of Gastrointestinal Cancer

Pallaval Veera Bramhachari and Nageswara Rao Reddy Neelapu

Abstract The current challenge for effective treatment of gastrointestinal cancer is detection of cancer at an early stage. Detection of gastrointestinal cancer at an early stage requires biomarkers expressing at early stage, biosensors, promising technologies, and computational methods. Therefore, this chapter discusses the role of biomarkers, biosensors, promising technologies, and computational methods which can be used for detection of gastrointestinal cancer. This provides information and new insights which can be used for early detection of gastrointestinal cancer.

Keywords Biomarkers · Biosensors · Computational methods for early detection of cancer · Technologies for early detection of cancer

1.1 Introduction

The current challenge for effective treatment of gastrointestinal cancer is detection of cancer at an early stage. Detection of gastrointestinal cancer at an early stage requires biomarkers expressing at early stage, biosensors, promising technologies, and computational methods (Fig. 1.1). Biomarkers provide understanding on features of cancer or tumor and helps in determining the features of cancer or tumor. Biosensors are used to sense or determine the biomarkers (expressed features of cancer or tumor) both qualitatively and quantitatively. The promising technologies are used either to identify biomarkers or develop sensors or devices. Computational methods used

P. Veera Bramhachari
Department of Biotechnology, Krishna University, Machilipatnam, Andhra Pradesh, India

N. R. R. Neelapu (✉)
Department of Biochemistry and Bioinformatics, GITAM Institute of Science, Gandhi Institute of Technology and Management (GITAM) (Deemed-to-be-University), Visakhapatnam, Andhra Pradesh, India

P. Veera Bramhachari, N. R. R. Neelapu (eds.), *Recent Advancements in Biomarkers and Early Detection of Gastrointestinal Cancers*, Diagnostics and Therapeutic Advances in GI Malignancies, https://doi.org/10.1007/978-981-15-4431-6_1

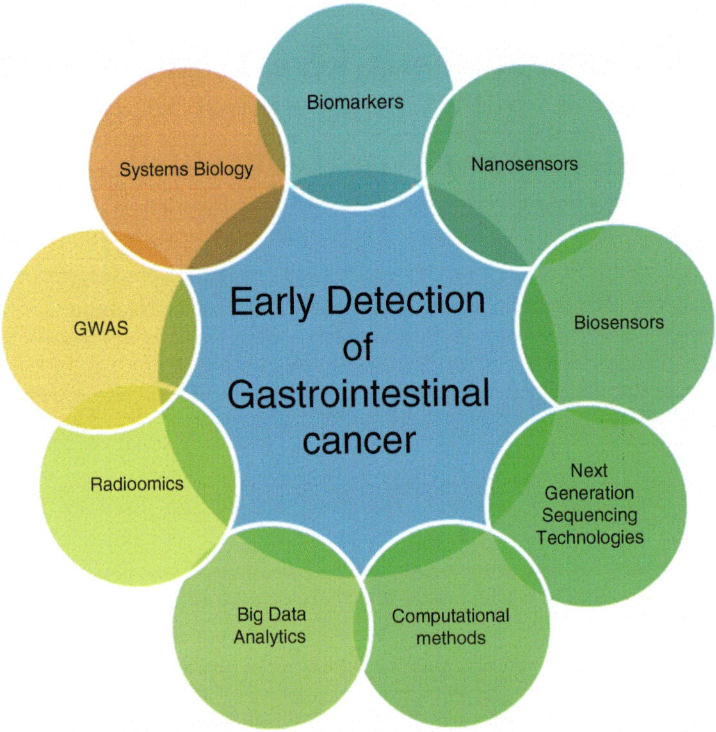

Fig. 1.1 Potential role of biomarkers, biosensors, technologies, and computational methods in early detection of gastrointestinal cancer

mathematical methods or expressions to identify or model or quantify features of cancer or tumor. Thus, this chapter discusses in detail about the role of biomarkers, biosensors, promising technologies, and computational methods which can be used for detection of gastrointestinal cancer. This provides information and new insights which can be used for early detection of gastrointestinal cancer.

1.2 Role of Biomarkers in Early Detection of Gastrointestinal Cancer

Molecular signatures or profiles generated by tumor or cancer due to proteins, microsatellite instability, hypermethylation, single nucleotide polymorphism, volatile compounds, serum, etc. are known as biomarkers. These markers are known as protein markers (if protein profiles are used), microsatellite instability markers (if microsatellite instability signatures are defined), hypermethylation markers (if hypermethylation profiles are identified), single nucleotide polymorphism markers (when SNP profiles are characterized), volatile markers (when volatile compounds are expressed), and serum markers (if serum profiles are outlined). These markers can be used to diagnose gastrointestinal cancer. Biomarkers are generally classified based on disease state, types of biomolecules, and other criteria (Radhika et al. 2016). The disease state biomarkers available for early detection of cancer are risk assessment biomarkers, screening/detection biomarkers, diagnosis biomarkers, prognosis biomarkers, prediction biomarkers, and monitoring biomarkers. The biomolecule biomarkers are DNA biomarker, RNA biomarker, protein biomarker, glycol biomarkers, metabolite biomarkers, and serum biomarkers (Radhika et al. 2016). The biomarkers in other criteria are imaging biomarkers, pathological biomarker, next-generation biomarkers, and in silico biomarkers (Radhika et al. 2016). Molecular markers like CDH1 gene (Bussemakers et al. 1994), DNMT3A gene (De Carvalho et al. 2012), PTPRCAP gene (Hyoungseok et al. 2009), PSCA gene (Sakamoto et al. 2008), VEGF-A gene (Yancopoulos et al. 2000), XRCC1 gene (Caldecott et al. 1996), IL-1 gene (England et al. 2014), HER-2 gene (Baselga et al. 1996), and MUC1 gene (Bafna et al. 2010) are known as genetically susceptible markers. These genetically susceptible markers are inherited by individual or population leading to cancer, which can be used to diagnose gastrointestinal cancer. The new dimension of cancer diagnosis is use of serum biomarkers in the development of serum biomarker panels which made diagnosis of gastrointestinal cancer simple based on serum profiles. Thus, it can be established that biomarkers have a role in the early diagnosis of gastrointestinal cancer.

1.3 Role of Biosensors in Early Detection of Gastrointestinal Cancer

Biosensors are used in fields like drug discovery (Morris 2013), fermentation industry (Yan et al. 2014), defense (Pohanka 2019), food quality (Torun et al. 2012), environmental monitoring (Arora et al. 2011), metabolic studies, and plant studies (Berens and Suess 2015). Biosensors can now provide key information on cancer for effective and safe treatment. Cost effectiveness, reliability, accuracy, and less time consuming are the important aspects of biosensors. DNA, antibody, antigen, enzyme, whole cell, and cell organelle are used as a biological recognition element for biosensors (Malhotra et al. 2017). The biological sample interacts with the element of the biosensor and forms a product (Malhotra et al. 2017). The product then reaches the transducer, amplifies, records, and displays on the devices (Malhotra et al. 2017). The different types of biosensors are affinity biosensor, catalytic biosensor, metabolism biosensor, DNA biosensor, electrochemical biosensor, optical biosensor, mass change biosensor, graphene-based biosensor, amperometric biosensor, microbial biosensor, miRNA biosensor, and many more (Leech 1994; Freitas et al. 2018; Jainish and Prittesh 2017; Medley et al. 2008; Tothill 2009; Kavita 2017; Lei et al. 2006; D'Souza 2001; Steinberg et al. 1995; Kumar et al. 2006; Choi and Chae 2012; Correia et al. 2017; Morgan et al. 2016; Rogers et al.

2016; Liu et al. 2017; Cheng et al. 2015; Kwon et al. 2018; Zhang et al. 2014; Szunerits and Boukherroub 2018). Biosensors role in early detection and diagnosis of cancer is known. Biosensors improved the diagnostic capability by its sensitivity, specificity, reproducibility, linearity, and high-throughput screening (Bhalla et al. 2016). Thus, biosensors have an important role in early detection of gastrointestinal cancer.

The second and the new dimension in biosensors is the use of nanotechnology for the development of biosensors. This shows how nanotechnology is a powerful and promising technology for early detection of gastrointestinal cancer. Nanobiosensors are developed using nanomaterial's like quantum dots, carbon nanotubes, nanopores, nanorods, nanowires, cantilevers, nanoparticles, and nanomembranes (Madani et al. 2013; de La Zerda and Gambhir 2007; Clarke et al. 2009; Hu et al. 2011; Israelsen et al. 2015; Zang et al. 2012; Daneshpour et al. 2016). Nanobiosensors role in early detection of cancer via carbon nanotube for CEA biomarker is reported (Länge et al. 2008). Nanobiosensors role in early detection of gastrointestinal cancer via electrochemical nanobiosensor for biomarkers miRNA 106A is also reported (Richardson et al. 2001). The different nanobiosensors are nanoparticles-based sensors (acoustic wave biosensors, magnetic biosensors, electrochemical biosensors), nanotube-based sensors, nanowire-based sensors, and ion channel-based sensors (Clark Jr and Lyons 1962; Desai et al. 1999; Cui et al. 2001; Cornell et al. 1997). The immobilization of biomolecules onto nanomaterials develops nanobiosensors for detection of analyte. The different strategies used for immobilization of biomolecules onto nanomaterials are covalent, noncovalent, and linker with covalent (Dubertret et al. 2002; Bruchez et al. 1998; Taton et al. 2001). The different parameters like selectivity, reproducibility, dynamic range, and negligible changes in concentrations of biomolecules indicate the performance of nanobiosensors. Thus, the potential role of nanobiosensors in early detection of gastrointestinal cancer can be materialized in the development of diagnostic devices or healthcare wearables in the near future.

1.4 Role of Technologies in Early Detection of Gastrointestinal Cancer

Expression profiling of microRNA, circulating microRNAs, serum microRNA, and plasma microRNA derive the signatures of the cancer. Expression profiling include RNA sequencing using next-generation sequencing technology or microarray of miRNAs using Affymetrix microarray or real-time reverse transcription PCR (qRT-PCR) to generate expression profile datasets. The analysis of these expression profile datasets between normal and gastric cancer cells would provide the differential expression profiles or patterns. The analysis indicates upregulated and downregulated miRNAs involved in signaling pathways related to environmental information processing and diseases. Therefore, these set of signature miRNAs may be promising biomarkers for the early diagnosis of gastrointesinal cancer.

In order to fulfill some of our knowledge gaps on cancer, it is essential to continually generate and explore omic's data on cancer. There are five next-generation sequencing technologies available to generate NGS data: first-, second-, third-, fourth-, and fifth-generation sequencing technologies. The first-generation sequencing technologies include Sanger sequencing and Maxam Gilbert sequencing method (Neelapu and Surekha 2016). The second-generation sequencing method includes Roche/454 Sequencing, Ion torrent sequencing, Illumina/Solexa sequencing, and ABI/SOLiD sequencing (Neelapu and Surekha 2016). The third-generation sequencing method includes Single Molecule Real-Time (SMRT) sequencing approach and Oxford Nanopore Technology (ONT) sequencing approach (Neelapu and Surekha 2016). The fourth-generation sequencing method includes Nanopore-based sequencing by biological nanopores and solid-state nanopores (Neelapu and Surekha 2016). The fifth-generation sequencing method includes high-fidelity nanopore sequencing of ultra-short DNA targets and cyclomics: ultra-sensitive nanopore sequencing of cell-free tumor DNA (Neelapu and Surekha 2016). The technologies like Roche/454 Sequencing, Illumina/Solexa sequencing, Single Molecule Real-Time (SMRT) sequencing approach and Nanopore-based sequencing are used to generate whole-genome sequencing (WGS) data (Neelapu and Surekha 2016). WGS of tumor or cancer cell followed by the analysis of WGS data provides genetic information and heterogeneity of tumor or cancer cell when compared with the normal cell (Nakagawa and Fujita 2018).

Whole exome sequencing (WES), is a genomic technique for sequencing all of the protein-coding regions of genes in a genome (known as the exome) (Ng et al. 2009). This information provides insights on understanding nature of tumor or cancer cell. Epigenome sequencing of cancer or tumor cell helps in understanding the epigenetic features regulating cancer cells or tumor. The technologies such as methylation-sensitive restriction enzyme sequencing (MRE-seq), methylated DNA immunoprecipitation sequencing (MeDIP-seq), methyl-CpG-binding domain protein sequencing (MBD-seq), reduced representation bisulfite sequencing (RRBS), whole-genome bisulfite sequencing (WGBS), oxidative bisulfite sequencing (oxBS-seq) generate epigenome data-based methylation patterns (Sarda and Hannenhalli 2014). The other technologies like chromatin immunoprecipitation sequencing (ChIP-seq), and chromatin immunoprecipitation-exonuclease (ChIP-exo) generate epigenome data based on histone modifications (Sarda and Hannenhalli 2014). These technologies helped in understanding epigenetic features regulating cancer cells or tumor. The deep sequencing of mRNA-seq, long-read, direct RNA-seq, and short sequence reads (transcriptomes) by RNA sequencing technologies helps in generating transcriptome data (Stark et al. 2019). The short-read cDNA, long-read c DNA, and long-read RNA are generated using platforms Illumina and Ion Torrent; PacBio and ONT; and Nanopore technology, respectively. This helps in understanding single-cell gene expression, translation (the translatome), RNA structure (the structurome), and spatial transcriptomics (spatialomics) and also aids in understanding nature of tumor or cancer cell (Stark et al. 2019). De novo peptide sequencing via tandem mass spectrometry is used to generate proteomics data (Dancík et al. 1999). This proteomics approach can be used to understanding nature of tumor or cancer cell. Thus, genomics, epigenomics, transcriptomics, and proteomics approach can be used understanding the nature of tumor or cancer cell.

1.5 Role of Computational Methods in Early Detection of Gastrointestinal Cancer

The role of computational methods like genome-wide association studies (GWAS) (Challa and Neelapu 2018), big data analytics, and systems biology approach is known and can be used for early detection of gastrointestinal cancer. The genome sequencing projects of human led to genome-wide association studies (GWAS) to recognize genes and its respective variants related with any traits or diseases. GWAS was used for prediction of early onset of gastrointestinal cancer and can be utilized as biomarker in the detection and prevention of gastrointestinal cancer. Biomarkers, like carcinoembryonic antigen (Länge et al. 2008) and carbohydrate antigen 19-9 (Perkins et al. 2003), are in clinical use for detection of advanced stage of gastric cancer. Genome studies reported that expression level of CDH1 (Suriano et al. 2003; Bacani et al. 2006), CTNNB1 (Zhou et al. 2002), CDX-2 (Seno et al. 2002; Mizoshita et al. 2003; Fan et al. 2005), HER2 (Moelans et al. 2011), CD44v6 (Carvalho et al. 2006), 5p15 (Du et al. 2013), PRKAA1 (Jiang et al. 2018), and Reprimo (Bernal et al. 2008) predictive biomarker for the early onset of gastric cancer.

Imaging biomarkers, pathological biomarker, next-generation biomarkers generate large amounts of data. Big data analytics help in analyzing the big data to discover diagnostics and therapeutics for gastric cancer. Radiomics is a process of converting digital medical images into mineable high-dimensional data, and radiomics uses machine-learning approach to make clinical decision (Lambin et al. 2012). Radiomics include correlating and integrating omics data with radiomics features extracted from radiological images and integrate them to create a more efficient and robust prognostic model (Lambin et al. 2012). Thus, radiomics helps in early detection of gastrointestinal cancer. In the same way, big data generated by imaging technologies, pathological methods or technologies, and next-generation sequencing technologies can be analyzed by employing various methods in big data analytics for early detection of cancer.

Systems biology approach integrated high-throughput and "omics" data in understanding the disease (Kang et al. 2016). Systems biology is necessary to analyze the complexities of various pathways involving signaling, regulation of the gene, cell metabolism, and alterations in its system caused due to mutations leading to malignancy (Kang et al. 2016). These approaches seem to be complicated with several interlinks connecting pathways, and it is necessary to signify it in the form of a computational model (Kang et al. 2016). And, also in identifying the proteins and pathways of gastric cancer that can be useful sequentially in identifying major proteins and pathways (Kang et al. 2016). This helps in understanding the functional difference that takes place from a normal and disease cell. Thus, systems biology approach helps in early detection of gastrointestinal cancer. However, these findings require further wet-lab validation.

1.6 Significance

Detection of gastrointestinal cancer at an early stage requires biomarkers expressing at early stage, biosensors, promising technologies, and computational methods. Next-generation sequencing technologies especially the fifth-generation technology like high-fidelity nanopore sequencing of ultra-short DNA targets and cyclomics: ultra-sensitive nanopore sequencing of cell-free tumor DNA can be used to identify biomarkers which are expressed at an early stage of cancer. NGS technologies can also be used to develop gene panels, whole exome sequencing (WES), and whole-genome sequencing for cancer detection. NGS gene panels are already in use for diagnosis of hereditary cancers like breast, ovarian, colon, etc. Biomarkers identified can be integrated into biosensors or nanosensors to develop real-time measurement devices for early detection of gastrointestinal cancer. Healthcare monitoring devices or healthcare wearables are already in market for diagnosing diseases like diabetes. These healthcare wearables may help in early detection of cancer, as well as help in monitoring the cancer patient condition and treatment outcome from time to time. The NGS technologies and omic's technologies may help in understanding the genes, epigenetic features, proteins, and other features responsible for transition of normal cell to cancer state. The computational methods help in developing new methods for analysis, and modeling of the data and also in mining the big data like radioimages. These computational methods may provide novel insights on cancer which can be used for detection of cancer. Thus, this chapter discusses about the potential role of biomarkers, biosensors, promising technologies, and computational methods for early detection of gastrointestinal cancer.

Acknowledgements NNRR is grateful to GITAM (Deemed-to-be-University) for providing necessary facilities to carry out the research work and for extending constant support. PVBC is thankful to Krishna University for providing necessary facilities to carry out the research work.

Conflict of Interest The authors declare that there is no potential conflict of interest.

References

Arora P, Sindhu A, Dilbaghi N, Chaudhury A (2011) Biosensors as innovative tools for the detection of food borne pathogens. Biosens Bioelectron 28(1):1–12

Bacani JT, Soares M, Zwingerman R, Bacani JT, Soares M, Zwingerman R, di Nicola N, Senz J, Riddell R, Huntsman DG, Gallinger S (2006) CDH1/E-cadherin germline mutations in early-onset gastric cancer. J Med Genet 43(11):867–872

Bafna S, Kaur S, Batra SK (2010) Membrane-bound mucins: the mechanistic basis for alterations in the growth and survival of cancer cells. Oncogene 29:2893–2904

Baselga J, Tripathy D, Mendelsohn J, Baughman S, Benz CC, Dantis L, Sklarin NT, Seidman AD, Hudis CA, Moore J, Rosen PP, Twaddell T, Henderson IC, Norton L (1996) Phase II study of weekly intravenous recombinant humanized anti-p185HER2 monoclonal antibody in patients with HER2/neu-overexpressing metastatic breast cancer. J Clin Oncol 14:737–744

Berens C, Suess B (2015) Riboswitch engineering—making the all-important second and third steps. Curr Opin Biotechnol 31:10–15

Bernal C, Aguayo F, Villarroel C, Bernal C, Aguayo F, Villarroel C, Vargas M, Díaz I, Ossandon FJ, Santibáñez E, Palma M, Aravena E, Barrientos C, Corvalan AH (2008) Reprimo as a potential biomarker for early detection in gastric cancer. Clin Cancer Res 14:6264–6269. https://doi.org/10.1158/1078-0432.CCR-07-4522

Bhalla N, Jolly P, Formisano N, Estrela P (2016) Introduction to biosensors. Essays Biochem 60 (1):1–8

Bruchez M, Moronne M, Gin P, Weiss S, Alivisatos AP (1998) Semiconductor nanocrystals as fluorescent biological labels. Science 281(5385):2013–2016

Bussemakers MJG, Van Bokhoven A, Völler M, Smit FP, Schalken JA (1994) The genes for the calcium-dependent cell adhesion molecules P- and E-cadherin are tandemly arranged in the human genome. Biochem Biophys Res Commun 203:291–1294

Caldecott KW, Aoufouchi S, Johnson P, Shall S (1996) XRCC1 polypeptide interacts with DNA polymerase beta and possibly poly (ADP-ribose) polymerase, and DNA ligase III is a novel molecular 'nick-sensor' *in vitro*. Nucleic Acids Res 24:4387–4394

Carvalho R, Milne ANA, Polak M, Offerhaus GJ, Weterman MA (2006) A novel region of amplification at 11p12-13 in gastric cancer, revealed by representational difference analysis, is associated with overexpression of CD44v6, especially in early-onset gastric carcinomas. Genes Chromosomes Cancer 45:967–975

Challa S, Neelapu NRR (2018) Genome-wide association studies (GWAS) for abiotic stress tolerance in plants. In: Biochemical, physiological and molecular avenues for combating abiotic. Academic, Cambridge, pp 135–150

Cheng F, Kardashliev T, Pitzler C, Shehzad A, Lue H, Bernhagen J et al (2015) A competitive flow cytometry screening system for directed evolution of therapeutic enzyme. ACS Synth Biol 4 (7):768–775

Choi S, Chae J (2012) An array of microliter-sized microbial fuel cells generating 100 μW of power. Sensors Actuators A Phys 177:10–15

Clark LC Jr, Lyons C (1962) Electrode systems for continuous monitoring in cardiovascular surgery. Ann N Y Acad Sci 102(1):29–45

Clarke J, Wu HC, Jayasinghe L, Patel A, Reid S, Bayley H (2009) Continuous base identification for single-molecule nanopore DNA sequencing. Nat Nanotechnol 4(4):265–270

Cornell BA, Braach-Maksvytis VLB, King LG, Osman PDJ, Raguse B, Wieczorek L, Pace RJ (1997) A biosensor that uses ion-channel switches. Nature 387(6633):580–583

Correia CN, Nalpas NC, McLoughlin KE, Browne JA, Gordon SV, MacHugh DE, Shaughnessy RG (2017) Circulating microRNAs as potential biomarkers of infectious disease. Front Immunol 8:118

Cui Y, Wei Q, Park H, Lieber CM (2001) Nanowire nanosensors for highly sensitive and selective detection of biological and chemical species. Science 293(5533):1289–1292

D'Souza SF (2001) Microbial biosensors. Biosens Bioelectron 16(6):337–353

Dancík V, Addona TA, Clauser KR, Vath JE, Pevzner PA (1999) De novo peptide sequencing via tandem mass spectrometry. J Comput Biol 6(3–4):327–342

Daneshpour M, Omidfar K, Ghanbarian H (2016) A novel electrochemical nanobiosensor for the ultrasensitive and specific detection of femtomolar level gastric cancer biomarker miRNA-106a. Beilstein J Nanotechnol 7(2016):2023–2036

De Carvalho DD, Sharma S, You JS, Su SF, Taberlay PC, Kelly TK, Yang X, Liang G, Jones PA (2012) DNA methylation screening identifies driver epigenetic events of cancer cell survival. Cancer Cell 21:655–667

de La Zerda A, Gambhir SS (2007) Drug delivery: keeping tabs on nanocarriers. Nat Nanotechnol 2 (12):745–746

Desai T, Hansford D, Ferrari M (1999) Characterization of micromachined silicon membranes for immunoisolation and bioseparation applications. J Membrane Sci 159:221–231

Du J, Xu Y, Dai J, Ren C, Zhu C, Dai N, Ma H, Shi Y, Hu Z, Lin D, Shen H, Jin G (2013) Genetic variants at 5p15 are associated with risk and early onset of gastric cancer in Chinese populations. Carcinogenesis 34:2539–2542

Dubertret B, Skourides P, Norris DJ, Noireaux V, Brivanlou AH, Libchaber A (2002) In vivo imaging of quantum dots encapsulated in phospholipid micelles. Science 298(5599):1759–1762

England H, Summersgill HR, Edye ME et al (2014) Release of interleukin-1α or interleukin-1β depends on mechanism of cell death. J Biol Chem 289(23):15942–15950

Fan Z, Li J, Dong B, Huang X (2005) Expression of Cdx2 and hepatocyte antigen in gastric carcinoma: correlation with histologic type and implications for prognosis. Clin Cancer Res 11:6162–6170

Freitas M, Nouws HP, Delerue-Matos C (2018) Electrochemical biosensing in cancer diagnostics and follow-up. Electroanalysis 30(8):1584–1603

Hu Y, Fine DH, Tasciotti E, Bouamrani A, Ferrari M (2011) Nanodevices in diagnostics. Wiley Interdiscip Rev Nanomed Nanobiotechnol 3(1):11–32

Hyoungseok JU, Lim B, Kim M, Kim YS, Kim WH, Ihm C, Noh S-M, Han DS, Yu H-J, Choi BY, Kang C (2009) A regulatory polymorphism at position −309 in PTPRCAP is associated with susceptibility to diffuse-type gastric cancer and gene expression. Neoplasia 11:1340–1347

Israelsen ND, Hanson C, Vargis E (2015) Nanoparticle properties and synthesis effects on surface-enhanced Raman scattering enhancement factor: an introduction. Sci World J 2015:124582. https://doi.org/10.1155/2015/124582

Jainish P, Prittesh P (2017) Biosensors and biomarkers: promising tools for cancer diagnosis. Int J Biosens Bioelectron 3(4):00072

Jiang Y, Li W, Lu J, Zhao X, Li L (2018) Association between PRKAA1 rs13361707 T>C polymorphism and gastric cancer risk: evidence based on a meta-analysis. Medicine (Baltimore) 97:e0302. https://doi.org/10.1097/MD.0000000000010302

Kang W, Cheng AS, Yu J, To KF (2016) Emerging role of Hippo pathway in gastric and other gastrointestinal cancers. World J Gastroenterol 22(3):1279

Kavita V (2017) DNA biosensors—a review. J Bioeng Biomed Sci 7:2

Kumar J, Jha SK, D'Souza SF (2006) Optical microbial biosensor for detection of methyl parathion pesticide using *Flavobacterium* sp. whole cells adsorbed on glass fiber filters as disposable biocomponent. Biosens Bioelectron 21(11):2100–2105

Kwon KK, Lee DH, Kim SJ, Choi SL, Rha E, Yeom SJ et al (2018) Evolution of enzymes with new specificity by high-throughput screening using DmpR-based genetic circuits and multiple flow cytometry rounds. Sci Rep 8(1):2659

Lambin P, Rios-Velazquez E, Leijenaar R, Carvalho S, van Stiphout RG, Granton P, Zegers CM, Gillies R, Boellard R, Dekker A, Aerts HJ (2012) Radiomics: extracting more information from medical images using advanced feature analysis. Eur J Cancer 48(4):441–446

Länge K, Rapp BE, Rapp M (2008) Surface acoustic wave biosensors: a review. Anal Bioanal Chem 391(5):1509–1519

Leech D (1994) Affinity biosensors. Chem Soc Rev 23(3):205–213

Lei Y, Chen W, Mulchandani A (2006) Microbial biosensors. Anal Chim Acta 568(1-2):200–210

Liu Y, Liu Y, Wang M (2017) Design, optimization and application of small molecule biosensor in metabolic engineering. Front Microbiol 8:2012

Madani SY, Shabani F, Dwek MV, Seifalian AM (2013) Conjugation of quantum dots on carbon nanotubes for medical diagnosis and treatment. Int J Nanomedicine 8:941–950
Malhotra S, Verma A, Tyagi N, Kumar V (2017) Biosensors: principle, types and applications. Int J Adv Res Innov Ideas Educ 3(2):3639–3644
Medley CD, Smith JE, Tang Z, Wu Y, Bamrungsap S, Tan W (2008) Gold nanoparticle-based colorimetric assay for the direct detection of cancerous cells. Anal Chem 80(4):1067–1072
Mizoshita T, Tsukamoto T, Nakanishi H, Inada K, Ogasawara N, Joh T, Itoh M, Yamamura Y, Tatematsu M (2003) Expression of Cdx2 and the phenotype of advanced gastric cancers: relationship with prognosis. J Cancer Res Clin Oncol 129:727–734
Moelans CB, Milne AN, Morsink FH, Offerhaus GJ, van Diest PJ (2011) Low frequency of HER2 amplification and overexpression in early onset gastric cancer. Cell Oncol (Dordr) 34:89–950
Morgan SA, Nadler DC, Yokoo R, Savage DF (2016) Biofuel metabolic engineering with biosensors. Curr Opin Chem Biol 35:150–158
Morris MC (2013) Fluorescent biosensors—probing protein kinase function in cancer and drug discovery. Biochim Biophys Acta 1834(7):1387–1395
Nakagawa H, Fujita M (2018) Whole genome sequencing analysis for cancer genomics and precision medicine. Cancer Sci 109(3):513–522
Neelapu NRR, Surekha C (2016) Next-generation sequencing and metagenomics. In: Wong K-C (ed) Computational biology and bioinformatics: gene regulation. CRC Press, Boca Raton, pp 331–351
Ng SB, Turner EH, Robertson PD, Flygare SD, Bigham AW, Lee C, Shaffer T, Wong M, Bhattacharjee A, Eichler EE, Bamshad M, Nickerson DA, Shendure J (2009) Targeted capture and massively parallel sequencing of 12 human exomes. Nature 461(7261):272–276
Perkins G, Slater E, Sanders G, Prichard J (2003) Serum tumor markers. Am Fam Physician 68 (6):1075–1082
Pohanka M (2019) Current trends in the biosensors for biological warfare agents assay. Materials (Basel) 12(14):2303. https://doi.org/10.3390/ma12142303
Radhika T, Jeddy N, Nithya S, Muthumeenakshi RM (2016) Salivary biomarkers in oral squamous cell carcinoma—an insight. J Oral Biol Craniofac Res 6(Suppl 1):S51–S54
Richardson J, Hawkins P, Luxton R (2001) The use of coated paramagnetic particles as a physical label in a magnetoimmunoassay. Biosens Bioelectron 16(9–12):989–993
Rogers JK, Taylor ND, Church GM (2016) Biosensor-based engineering of biosynthetic pathways. Curr Opin Biotechnol 42:84–91
Sakamoto H, Yoshimura K, Saeki N, Katai H, Shimoda T, Matsuno Y, Saito D, Sugimura H, Tanioka F, Kato S, Matsukura N, Matsuda N, Nakamura T, Hyodo I, Nishina T, Yasui W, Hirose H, Hayashi M, Toshiro E, Ohnami S, Sekine A, Sato Y, Totsuka H, Ando M, Takemura R, Takahashi Y, Ohdaira M, Aoki K, Honmyo I, Chiku S, Aoyagi K, Sasaki H, Ohnami S, Yanagihara K, Yoon KA, Kook MC, Lee YS, Park SR, Kim CG, Choi IJ, Yoshida T, Nakamura Y, Hirohashi S (2008) Genetic variation in PSCA is associated with susceptibility to diffusetype gastric cancer. Nat Genet 40:730–740
Sarda S, Hannenhalli S (2014) Next-generation sequencing and epigenomics research: a hammer in search of nails. Genomics Inform 12(1):2–11. https://doi.org/10.5808/GI.2014.12.1.2
Seno H, Oshima M, Taniguchi M-A, Usami K, Ishikawa TO, Chiba T, Taketo MM (2002) CDX2 expression in the stomach with intestinal metaplasia and intestinal-type cancer: prognostic implications. Int J Oncol 21:769–774
Stark R, Grzelak M, Hadfield J (2019) RNA sequencing: the teenage years. Nat Rev Genet 20:631–656
Steinberg SM, Poziomek EJ, Engelmann WH, Rogers KR (1995) A review of environmental applications of bioluminescence measurements. Chemosphere 30(11):2155–2197
Suriano G, Oliveira C, Ferreira P, Machado JC, Bordin MC, De Wever O, Bruyneel EA, Moguilevsky N, Grehan N, Porter TR, Richards FM, Hruban RH, Roviello F, Huntsman D, Mareel M, Carneiro F, Caldas C, Seruca R (2003) Identification of CDH1 germline missense

mutations associated with functional inactivation of the E-cadherin protein in young gastric cancer probands. Hum Mol Genet 12:575–582

Szunerits S, Boukherroub R (2018) Graphene-based biosensors. Interface Focus 8(3):20160132

Taton TA, Lu G, Mirkin CA (2001) Two-color labeling of oligonucleotide arrays via size-selective scattering of nanoparticle probes. J Am Chem Soc 123(21):5164–5165

Torun Ö, Boyacı İH, Temür E, Tamer U (2012) Comparison of sensing strategies in SPR biosensor for rapid and sensitive enumeration of bacteria. Biosens Bioelectron 37(1):53–60

Tothill IE (2009) Biosensors for cancer markers diagnosis. Semin Cell Dev Biol 20(1):55–62

Yan C, Dong F, Chun-yuan B, Si-rong Z, Jian-guo S (2014) Recent progress of commercially available biosensors in China and their applications in fermentation processes. J Northeast Agric Univ 21(4):73–85

Yancopoulos GD, Davis S, Gale NW, Rudge JS, Wiegand SJ, Holash J (2000) Vascular-specific growth factors and blood vessel formation. Nature 407:242–248

Zang D, Ge L, Yan M, Song X, Yu J (2012) Electrochemical immunoassay on a 3D microfluidic paper-based device. Chem Commun 48:4683–4685

Zhang Y, Gao G, Liu H, Fu H, Fan J, Wang K, Chen Y, Li B, Zhang C, Zhi X, He L (2014) Identification of volatile biomarkers of gastric cancer cells and ultrasensitive electrochemical detection based on sensing interface of Au-Ag alloy coated MWCNTs. Theranostics 4(2):154

Zhou YN, Xu CP, Han B, Li M, Qiao L, Fang DC, Yang JM (2002) Expression of E-cadherin and β-catenin in gastric carcinoma and its correlation with the clinicopathological features and patient survival. World J Gastroenterol 8:987–993

Chapter 2
Biomarkers as the Promising Tools for Early Detection of Gastrointestinal Cancer

Pallaval Veera Bramhachari and Nageswara Rao Reddy Neelapu

Abstract Gastrointestinal cancer is one of the most prevalent types in the world and which is poorly understood at the molecular level. Early detection of gastric cancer is still a problem and detection of cancer at an early stage will help plan the selection of an appropriate treatment plan and effective monitoring of diseases. Literature reports the use of biomarkers and methods for early detection of cancer. This chapter summarizes the burden of cancer, especially gastrointestinal cancer, methods for diagnosis of cancer, the importance of biomarkers and early detection of cancer, biomarkers and their classification, and biomarkers for gastrointestinal cancers which could be potentially used for early diagnosis, and accurate prediction of therapeutic approaches.

Keywords Biomarkers · Burden of cancer · Cancer · Early detection of cancer · Gastrointestinal cancer · Technologies for early detection of cancer

2.1 Introduction

Global cancer burden shows that 43.8 million people are living with cancer (GLOBOCAN Database 2018). The new global cancer burden in 2018 is 18.1 million new cases, whereas the lethality of cancer is 9.6 million cancer deaths in the year 2018 (GLOBOCAN Database 2018). The burden of cancer is different in different regions, where 50% of the cancer cases were registered in Asia, 25% cases accounted to Europe, and the rest of the cases are distributed across the different parts of the world (GLOBOCAN Database 2018). Gastrointestinal cancer (GI) is the

P. Veera Bramhachari
Department of Biotechnology, Krishna University, Machilipatnam, Andhra Pradesh, India

N. R. R. Neelapu (✉)
Department of Biochemistry and Bioinformatics, GITAM Institute of Science, Gandhi Institute of Technology and Management (GITAM) (Deemed-to-be-University), Visakhapatnam, Andhra Pradesh, India

P. Veera Bramhachari, N. R. R. Neelapu (eds.), *Recent Advancements in Biomarkers and Early Detection of Gastrointestinal Cancers*, Diagnostics and Therapeutic Advances in GI Malignancies, https://doi.org/10.1007/978-981-15-4431-6_2

third-highest based on lethality and fourth-highest based on morbidity of all cancers. The gastrointestinal cancer burden can be assumed from the new and death cases registered. Nearly ~1,033,701 new cases and ~782,000 deaths were recorded for gastrointestinal cancer in 2018 (GLOBOCAN Database 2018). Statistical techniques like Bayesian inference methods, capture-recapture methods, Mortality and Incidence Analysis Model (MIAMOD), and Prevalence and Incidence Analysis Model (PIAMOD) are the methods used to measure the burden of cancer on population (Sharifian et al. 2016). This chapter discusses in detail how the diagnosis and early detection of cancer can relieve the burden of cancer. Also, the details of biomarkers and their types that can be used for early diagnosis of gastric cancer are discussed. This helps in understanding the role of biomarkers for early diagnosis of gastric cancer.

2.2 Diagnosis and Importance of Early Detection of Cancer

Diagnosis is a very important aspect to confirm the disease, especially cancer. The diagnosis of cancer provides an opportunity to treat the diseases appropriately. But, by the time the patient is diagnosed with cancer; the patient is in an advanced stage of cancer with his life at risk. To save their lives, cancer patients can be diagnosed early. Therefore, this section provides details on the diagnosis of cancer and the importance of early detection of cancer.

2.2.1 Diagnosis of Cancer

Broadly, there are four types of tests like tumor testing for biomarkers, cytogenetic tests, gene tests, and biochemical tests available for diagnosis of cancers. This section discusses briefly the tests for diagnosing cancers or tumors.

2.2.1.1 Tumor Testing for Biomarkers

Samples of blood, body tissue, bodily fluids, tissue biopsies, and urine are used for testing tumor biomarkers. Molecular or genetic tests identify molecular features of genes or DNA in cells of cancer or tumor. These molecular features are specific biomarkers for cancer. PCA3 and T2: ERG are the biomarkers of prostate cancer identified by gene testing (Füzéry et al. 2013; Paddock 2019).

2.2.1.2 Gene Tests

Molecular tests look for biomarkers like genes (inside chromosomes), extra copies of a gene (duplicated or amplified genes), missing genes (gene deletions), incorrectly placed genes (translocated genes), changes in genes (mutated genes) in small tissue samples, blood tests, liquid biopsies, and biopsy (tissue testing). Specific biomarkers like *HER2* or *EGFR* (single-gene test) or gene-expression panels for many biomarkers are the molecular tests used for the diagnosis of cancer (Chanley 2018).

2.2.1.3 Cytogenetic Tests

The structural abnormalities in chromosomes leading to cancer can be diagnosed with cytogenetic tests. Samples of blood cells, tissues, and bone marrow can be used to measure the abnormalities in the chromosome. The specific changes in the chromosomes can act as biomarkers to screen or diagnose cancer. A change in the Philadelphia chromosome is the biomarker and the common feature of blood cancer (chronic myeloid leukemia) (Chanley 2018).

2.2.1.4 Biochemical Tests

Mutated genes express abnormal proteins and biochemical tests identify these proteins which serve as biomarkers. For example, the gene test uses the HER2 gene, whereas biochemical tests look for HER2 protein in the tissue sample. The tests described above identify a biomarker in the cancer cells and help in characterizing the specific nature of cancer. Understanding biomarkers related to cancer may help to get the best treatment for cancer and also show whether cancer is responding to treatment or not (Chanley 2018).

2.2.2 *Importance of Early Detection of Cancer*

The importance of early detection of cancer can only be addressed when it is understood, why some cancers are diagnosed late? how finding and treating cancer at an early stage can save lives? and how early diagnosis can improve survival? Some cancers are diagnosed late and the reason for the delay in cancer diagnosis is low awareness of cancer signs and symptoms among the general public, health care providers, physicians, and nurses (Why is early diagnosis important 2019). The signs and symptoms of cancer can be abnormal bleeding, chronic hoarseness, lumps, persistent indigestion, and sores that fail to heal. Education promoting sessions on cancer signs and symptoms would create awareness and encourage screening or early diagnosis of cancer. Early detection of cancer greatly increases the chances for successful treatment, whereas if cancer is diagnosed late then treatment becomes more difficult, decreasing the chances of survival of the patient. Early diagnosis is particularly relevant for cancers of the breast, cervix, mouth, larynx, colon and rectum, and skin. Some predictions estimated, how early diagnosis can improve the survival of cancer patients. In the case of bowel cancers, nine of the ten patients can survive if diagnosed at an early stage (Why is early diagnosis important 2019). In the case of breast and ovarian cancer, 90% of women survive for more than 5 years if diagnosed at an early stage when compared with women who are diagnosed at an advanced stage (Why is early diagnosis important 2019). In the case of lung cancer, 80% of patients survive for a year if diagnosed at an early stage when compared with patients who are diagnosed at an advanced stage (Why is early diagnosis important 2019). These advantages demonstrate the importance of early diagnosis of cancer. The role of biomarkers in early diagnosis of cancer is well known and established. Further, details on the importance of biomarkers would provide an understanding of the early diagnosis of cancer.

2.3 Importance of Biomarkers

Biomarkers are the molecules produced by the tumor or cancer cells in the body due to changes in genes, DNA or RNA that transform the normal properties of the cell and develop the cell into cancer cells. Biomarkers can determine the characteristics of a tumor or cancer, and also the severity or grade of cancer. Understanding the characteristics of a tumor or cancer allows physicians to customize treatment to cancer. This paved path and revolutionized the treatment of cancer by approaches like personalized medicine or precision medicine. Hence, cancer or tumor biomarkers can be identified through gene or molecular testing and can be characterized to understand the paint a specific picture of a tumor. Once the biomarker is recognized, targeted therapy can be designed for specific cancer with reduced cost and side effects.

2.4 Biomarkers Available for Early Detection of Cancer

Biomarkers are classified based on disease state, types of biomolecules, and other criteria (Radhika et al. 2016) (Fig. 2.1).

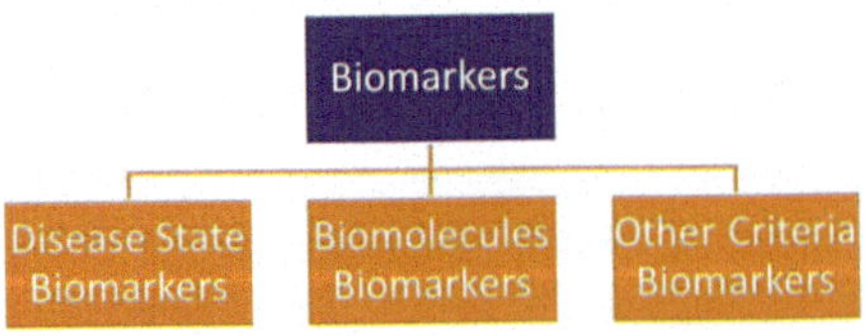

Fig. 2.1 Biomarkers and classification of biomarkers

2.4.1 Disease State Biomarkers

The disease state biomarkers available for early detection of cancer include risk assessment biomarkers, screening/detection biomarkers, diagnosis biomarkers, prognosis biomarkers, prediction biomarkers, and monitoring biomarkers (Radhika et al. 2016). Risk assessment biomarkers are associated with detecting the risk concerning predisposition of gene mutations in individuals which can lead to cancer. Risk assessment biomarkers can help in identifying the risk of cancer at an early stage (Radhika et al. 2016). Screening or detection biomarkers are real-time indicators like antibodies, serum proteins, circulating tumor cells, and DNA fragments in the bloodstream reflecting cancer or tumor. These indicators or biomarkers help in screening or detecting cancer or tumor (Radhika et al. 2016). Diagnosis biomarkers can determine, confirm the primary origin of cancer or tumor in the biopsy sample (Radhika et al. 2016). Prognosis biomarkers provide information about a patient's expected outcome, regardless of therapy. Sometimes, cancers are more aggressive than others and prognosis biomarkers can help in determining which cancers may grow rapidly and/or metastasize (Radhika et al. 2016). Prediction biomarkers are used to predict a patient's response to the drug and its dose when used for cancer treatment. Cancer is a heterogeneous disease, and different cancers respond differently to the same treatment methods and prediction biomarkers are used to predict a patient's response to the treatment (Radhika et al. 2016). Monitoring biomarkers are used to predict and monitor a patient's cancer recurrence after treatment. Thus, risk assessment biomarkers, screening/detection biomarkers, diagnosis biomarkers, prognosis biomarkers, prediction biomarkers, and monitoring biomarkers are available for early detection of cancer (Fig. 2.2). Biomarkers used for early detection of gastrointestinal cancer are listed below in Table 2.1. These biomarkers can be employed by different technologies available for early detection of gastrointestinal cancer.

2.4.2 Biomolecule Biomarkers

The biomolecule biomarkers include DNA biomarkers, RNA biomarkers, protein biomarkers, glycol biomarkers, metabolite biomarkers, and serum biomarkers (Radhika et al. 2016) (Fig. 2.3). Certain races or populations are susceptible to cancer, who are either predisposed or acquire genetic material hereditarily via DNA. These are known as biomarkers of genetic susceptibility or DNA biomarkers (Biomarkers Definitions Working Group 2001). MicroRNAs, circulating microRNAs, and plasma microRNAs are a few examples of RNA biomarkers. MicroRNAs are endogenous single-stranded non-coding small RNA molecules that are secreted into the circulation and exist stably. These, microRNAs exhibit aberrant expression under different physiological and pathological conditions. These differentially expressed circulating microRNAs are the potential biomarkers for cancer screening (Wang

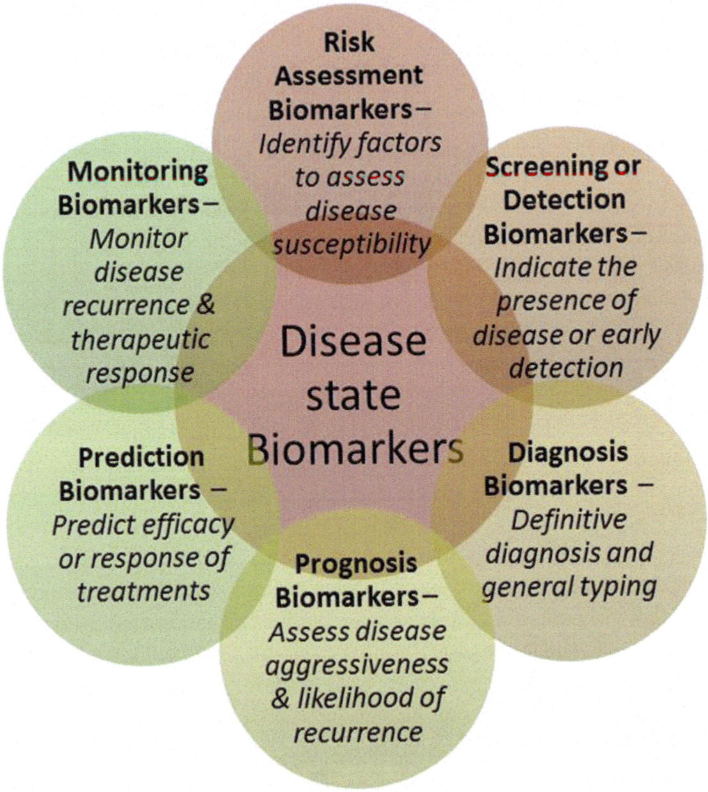

Fig. 2.2 Disease biomarkers available for early detection of cancer

et al. 2018). Circulating microRNAs in the serum generating microRNAs expression profiles are known as serum microRNAs (Wang et al. 2013). Circulating microRNAs when isolated from plasma of human subjects and generate expression profiles on microRNAs are known as plasma microRNAs (Wozniak et al. 2015). These, circulating microRNAs have several clinical applications like a diagnosis of cancer, classification of the tumor, monitoring, and outcome prognosis. Proteins causing disease or associated with susceptibility of the disease are known as protein biomarkers (Biomarkers Definitions Working Group 2001). Immunoassays and mass spectrometry assays are the two types of protein biomarkers assay platforms available for the discovery of protein biomarkers (Walid and Klaus 2010). These protein biomarkers have several clinical applications like a diagnosis of cancer and the classification of the tumor. Reactive oxygen species like hydroxyl radicals (HO∗) have generated which damage DNA, i.e., thymidine during oxidation forming thymidine glycol (5,6-dihydroxy-5,6-dihydrothymidine). Thymidine glycol is a biomarker that is excreted via urine and can be estimated as a biomarker for its disease state (Makropoulos et al. 2000). Volatile or metabolite biomarkers are

Table 2.1 List of biomarkers for gastrointestinal cancer

S. no	Biomarkers	References
1	HER2 (ERBB2), EGFR, VEGFA, NOTCH1, p-mTOR, MMP1, MMP7, TGFB1, MET, HER3 (ERBB3), SHH/PTCH1/SMO, FGFR2, CASOX9, TP53, PTEN, ALDH, PIK3	Elimova et al. (2015)
2	PD-L1	Curea et al. (2017)
3	ADAM23, GDNF, MINT25, MLF1, PRDM5, RORA	Watanabe et al. (2009)
4	BARHL2	Yamamoto et al. (2016)
5	PVT1	Yuan et al. (2016)
6	CagA	Saju et al. (2016)
7	VacA	Ghotaslou et al. (2018)
8	Gastrokine 1	Altieri et al. (2017)
9	CEACEM6, APOC1, YF13H12, CDH17, FUS, COLIA1, COLIA2, APOE	Yasui et al. (2004)
10	OLFM4, HOXA10, DSC2, TSPAN8, TM9SF3	Oue et al. (2015)
11	CCNB1 and CCNB2	Wang et al. (2015)
12	ZNF331, ZSCAN18, CDO1	Marie Vedeld et al. (2015)
13	KLK6	Paliouras et al. (2007)

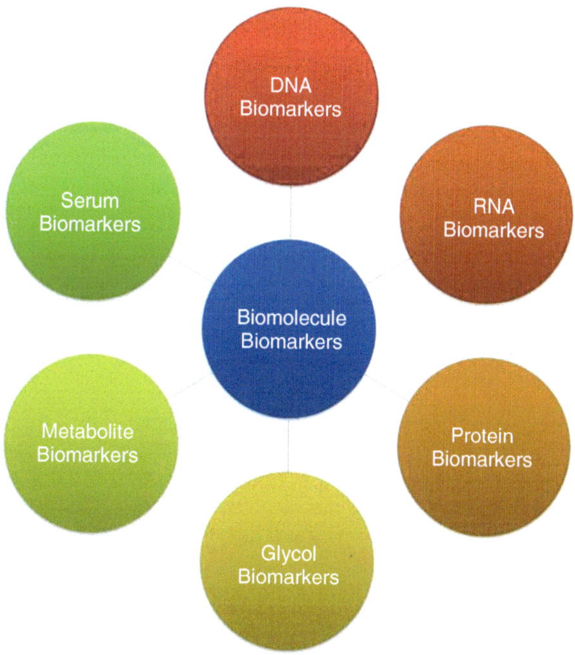

Fig. 2.3 Biomolecule biomarkers and their classification

volatile organic compounds (VOCs) released from human body fluids by endogenous metabolic processes. Expressions of VOCs bring in pathophysiological changes leading to disease, and several disease-specific volatile biomarkers have been identified and used in diagnostic aids (Kwak and Preti 2011). Serum biomarkers are substances synthesized by the tumor or cancer cells and released into circulation or expressed at the cell surface in large quantity changing quantitatively the serum during tumor or cancer development (Kato and Torigoe 1977).

2.4.3 Other Criteria Biomarkers

The biomarkers in other criteria include imaging biomarkers, pathological biomarkers, next-generation biomarkers, and in silico biomarkers (Radhika et al. 2016) (Fig. 2.4). Biologic feature of an image measured using techniques like CT, electroencephalography, magnetoencephalography, MRI to diagnose patients is known as imaging biomarker (Smith et al. 2003). Histopathologic techniques like electron microscopy, confocal laser scanning microscopy, immunohistochemistry, and in situ hybridization detect morphology of the disease state and improve diagnoses. These morphological and pathological features are known as pathological biomarkers (Novilla et al. 2014). Markers that are generated/identified using next-generation technologies like pharmacogenetics/pharmacogenomics, genotypic drug metabolism and transport, haplotype and SNP, RNA expression profiling, metabolomics, proteomics for the clinical outcomes during the development program are called next-generation biomarkers (Hogan et al. 2018). The different types of biomarkers are—pharmacogenetic biomarkers, pharmacogenomics biomarkers, genotypic drug metabolism biomarkers, drug transport biomarkers, haplotype biomarkers, SNPs, RNA expression profiles, metabonomics biomarkers, and proteomics biomarkers (Hogan et al. 2018). Computational or in silico methodologies are used to detect the pathological changes and connectivity of cells especially in

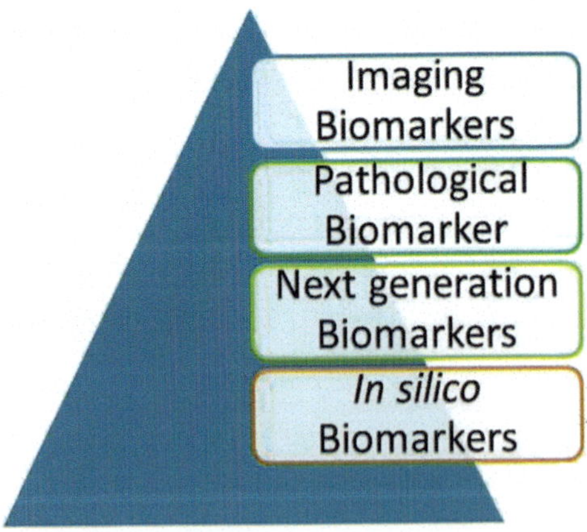

Fig. 2.4 Biomarkers in other criteria and their classification

neurons or any other tissues. These biomarkers are known as computational or in silico biomarkers (Siekmeier 2017). These biomarkers can be employed by different technologies available for early detection of gastrointestinal cancer. The different technologies available for early detection of cancer are DNA sequencing, next-generation sequencing technologies, "omics" technologies, nanotechnology, synthetic biology, next-generation sequencing panels (exomes to genomes), serum biomarker panels, ultra-sensitive nano-chips, nanosensors, nanodevices, biosensors, electrochemical biosensors, DNA biosensors, synthetic biology devices, etc.

2.5 Conclusions and Future Perspectives

Gastrointestinal cancer is one of the most prevalent, ranking third highest based on lethality and fourth-highest based on morbidity. Tumor testing for biomarkers, cytogenetic tests, gene tests, and biochemical tests are the tests available for the diagnosis of cancers. Early detection of gastric cancer is still a problem and the reasons for the delay in cancer diagnosis are low awareness of cancer signs and symptoms among the general public, health care providers, physicians, and nurses. The signs and symptoms of cancer are abnormal bleeding, chronic hoarseness, lumps, persistent indigestion, and sores that fail to heal. Promoting education and awareness of cancer signs and symptoms would encourage screening or early diagnosis of cancer. Early detection of cancer greatly increases the chances for successful treatment, whereas if cancer is diagnosed late then treatment becomes more difficult, decreasing the chances of survival of the patient. Early diagnosis is particularly relevant for cancers of the breast, cervix, mouth, larynx, colon and rectum, and skin. Biomarkers are used for early detection of cancer and biomarkers are classified based on disease state, types of biomolecules, and other criteria. The disease state biomarkers available for early detection of cancer are risk assessment biomarkers, screening/detection biomarkers, diagnosis biomarkers, prognosis biomarkers, prediction biomarkers, and monitoring biomarkers. The biomolecule biomarkers are DNA biomarkers, RNA biomarkers, protein biomarkers, glycol biomarkers, metabolite biomarkers, and serum biomarkers. The biomarkers in other criteria are imaging biomarkers, pathological biomarkers, next-generation biomarkers, and in silico biomarkers. Detection of cancer at an early stage using biomarkers will help to plan the selection of an appropriate treatment plan and effective monitoring of diseases.

Acknowledgments NNRR is grateful to GITAM (Deemed-to-be-University) for providing necessary facilities to carry out the research work and for extending constant support. PVBC is thankful to Krishna University for providing the necessary facilities to carry out the research work.

Conflict of Interest The authors declare that there is no potential conflict of interest.

References

Altieri F, Di Stadio CS, Federico A, Miselli G, De Palma M, Rippa E, Arcari P (2017) Epigenetic alterations of gastrokine 1 gene expression in gastric cancer. Oncotarget 8(10):16899

Biomarkers Definitions Working Group (2001) Biomarkers and surrogate endpoints: preferred definitions and conceptual framework. Clin Pharmacol Ther 69(3):89–95

Chanley K (2018) What are cancer biomarkers, and why re they important? https://conquer-magazine.com/issues/special-issues/may-2018-cancer-biomarkers-and-molecular-testing/682-what-are-cancer-biomarkers-and-why-are-they-important. Accessed 4 Oct 2019

Curea FG, Hebbar M, Ilie SM, Bacinschi XE, Trifanescu OG, Botnariuc I, Anghel RM (2017) Current targeted therapies in HER2-positive gastric adenocarcinoma. Cancer Biother Radiopharm 32(10):351–363

Elimova E, Wadhwa R, Shiozaki H, Sudo K, Estrella JS, Badgwell BD, Ajani JA (2015) Molecular biomarkers in gastric cancer. J Natl Compr Cancer Netw 13(4):e19–e29

Füzéry AK, Levin J, Chan MM, Chan DW (2013) Translation of proteomic biomarkers into FDA approved cancer diagnostics: issues and challenges. Clin Proteomics 10(1):13

Ghotaslou R, Leylabadlo HE, Nasiri MJ, Dabiri H, Hashemi A (2018) Risk of gastric cancer in association with Helicobacter pylori different virulence factors: a systematic review and meta-analysis. Microb Pathog 118:214–219

GLOBOCAN Database (2018). http://gco.iarc.fr/today/. Accessed October 2019

Hogan SA, Levesque MP, Cheng PF (2018) Melanoma immunotherapy: next-generation biomarkers. Front Oncol 8:178. https://doi.org/10.3389/fonc.2018.00178

Kato H, Torigoe T (1977) Radioimmunoassay for tumor antigen of human cervical squamous cell carcinoma. Cancer 40:1621–1628

Kwak J, Preti G (2011) Volatile disease biomarkers in breath: a critique. Curr Pharm Biotechnol 12 (7):1067–1074

Makropoulos W, Kocher K, Heintz B, Schwarz ER, Mertens PR, Stefanidis I (2000) Urinary thymidine glycol as a biomarker for oxidative stress after kidney transplantation. Ren Fail 22 (4):499–510

Marie Vedeld H, Andresen K, Andrassy Eilertsen I, Nesbakken A, Seruca R, Gladhaug IP et al (2015) The novel colorectal cancer biomarkers CDO1, ZSCAN18 and ZNF331 are frequently methylated across gastrointestinal cancers. Int J Cancer 136(4):844–853

Novilla MN, Meador VP, Jacobson SB (2014) Pathological biomarkers in toxicology in biomarkers in toxicology. In Gupta RC (ed), Academic, Cambridge, pp 907–919

Oue N, Sentani K, Sakamoto N, Yasui W (2015) Clinicopathologic and molecular characteristics of gastric cancer showing gastric and intestinal mucin phenotype. Cancer Sci 106(8):951–958

Paddock C (2019) New prostate cancer blood test could reduce biopsies. https://www.medicalnewstoday.com/articles/326366.php. Accessed 4 Oct 2019

Paliouras M, Borgono C, Diamandis EP (2007) Human tissue kallikreins: the cancer biomarker family. Cancer Lett 249(1):61–79

Radhika T, Jeddy N, Nithya S, Muthumeenakshi RM (2016) Salivary biomarkers in oral squamous cell carcinoma—an insight. J Oral Biol Craniofac Res 6(Suppl 1):S51–S54

Saju P, Murata-Kamiya N, Hayashi T, Senda Y, Nagase L, Noda S, Seto Y (2016) Host SHP1 phosphatase antagonizes Helicobacter pylori CagA and can be downregulated by Epstein–Barr virus. Nat Microbiol 1(4):16026

Sharifian A, Pourhoseingholi MA, Baghestani A, Hajizadeh N, Gholizadeh S (2016) Burden of gastrointestinal cancers and problem of the incomplete information; how to make up the data? Gastroenterol Hepatol Bed Bench 9(1):12–17

Siekmeier PJ (2017) An insilico, biomarker-based method for the evaluation of virtual neuropsychiatric drug effects. Neural Comput 29(4):1021–1052

Smith JJ, Sorensen AG, Thrall JH (2003) Biomarkers in imaging: realizing radiology's future. Radiology 227(3):633–638

Walid QM, Klaus L (2010) Protein biomarker immunoassays: opportunities and challenges. https://www.ddw-online.com/precision-medicine/p142790-protein-biomarker-immunoassays:-opportunities-and-challenges.html. Accessed 4 Oct 2019

Wang WT, Zhao YN, Han BW, Hong SJ, Chen YQ (2013) Circulating microRNAs identified in a genome-wide serum microRNA expression analysis as noninvasive biomarkers for endometriosis. J Clin Endocrinol Metab 98(1):281–289

Wang DG, Chen G, Wen XY, Wang D, Cheng ZH, Sun SQ (2015) Identification of biomarkers for diagnosis of gastric cancer by bioinformatics. Asian Pac J Cancer Prev 16(4):1361–1365

Wang H, Peng R, Wang J, Qin Z, Xue L (2018) Circulating microRNAs as potential cancer biomarkers: the advantage and disadvantage. Clin Epigenetics 10:59. https://doi.org/10.1186/s13148-018-0492-1

Watanabe Y, Kim HS, Castoro RJ, Chung W, Estecio MR, Kondo K, Suk KT (2009) Sensitive and specific detection of early gastric cancer with DNA methylation analysis of gastric washes. Gastroenterology 136(7):2149–2158

Why is early diagnosis important? https://www.cancerresearchuk.org/about-cancer/cancer-symptoms/why-is-early-diagnosis-important. Accessed 4 Oct 2019

Wozniak MB, Scelo G, Muller DC, Mukeria A, Zaridze D, Brennan P (2015) Circulating microRNAs as non-invasive biomarkers for early detection of non-small-cell lung cancer. PLoS One 10(5):e0125026. https://doi.org/10.1371/journal.pone.0125026

Yamamoto H, Watanabe Y, Oikawa R, Morita R, Yoshida Y, Maehata T, Itoh F (2016) BARHL2 methylation using gastric wash DNA or gastric juice exosomal DNA is a useful marker for early detection of gastric cancer in an *H. pylori*-independent manner. Clin Transl Gastroenterol 7(7):e184

Yasui W, Oue N, Ito R, Kuraoka K, Nakayama H (2004) Search for new biomarkers of gastric cancer through serial analysis of gene expression and its clinical implications. Cancer Sci 95(5):385–392

Yuan CL, Li H, Zhu L, Liu Z, Zhou J, Shu Y (2016) Aberrant expression of long noncoding RNA PVT1 and its diagnostic and prognostic significance in patients with gastric cancer. Neoplasma 63(3):442–449

Chapter 3
Development and Evaluation of Biomarkers for Early Detection of Cancer

Surekha Challa, Ravi Chandra Pavan Kumar Sri-Tirumala-Peddiniti, and Nageswara Rao Reddy Neelapu

Abstract Biomarkers are indicators of normal biological or pathogenic processes or pharmacological responses to a therapeutic intervention. In the last few years, pharmaceutical companies have shown interest in biomarkers. They incorporated biomarkers into company's drug development program and used them as companion tests for targeted therapeutics. The pharmaceutical industries have witnessed a dramatic increase in the biomarkers market in the last 7 years. The volume of business generated demonstrates demarcated IP filings in the last 5 years. Biomarkers' discovery and development is a complex process, which involves research, assay development, and commercialization. The research includes discovery and identification of biomarkers using different methods, whereas assay development includes exploratory phase, probable valid phase and known valid phase, following the commercialization of biomarkers with FDA approval. This chapter discusses in detail about types of biomarkers, the volume of biomarkers business, discovery and development of biomarkers, and applications of biomarkers.

Keywords Biomarkers · Discovery of biomarkers · Assay of biomarkers · Development of biomarkers · Commercialization of biomarkers

3.1 Introduction

Biomarkers' role in present science is remarkably increasing day by day in making decisions in every phase of drug discovery and development, and also in disease diagnosis. In the initial phases of drug development, biomarkers were used to evaluate their activity in animal models to prove their mode of action. In the later stages of drug development, biomarkers are used to make decisions in the evaluation

---

S. Challa · R. C. P. K. Sri-Tirumala-Peddiniti · N. R. R. Neelapu (✉)
Department of Biochemistry and Bioinformatics, GITAM Institute of Science, Gandhi Institute of Technology and Management (GITAM) (Deemed-to-be-University), Visakhapatnam, Andhra Pradesh, India

P. Veera Bramhachari, N. R. R. Neelapu (eds.), *Recent Advancements in Biomarkers and Early Detection of Gastrointestinal Cancers*, Diagnostics and Therapeutic Advances in GI Malignancies, https://doi.org/10.1007/978-981-15-4431-6_3

of pharmacologic effect and safety in animal models and humans by which we can predict the patient's response to the compound. Developments on biomarkers by pharma industries made easy to study patient's genetic makeup as well as medical profile to receive personalized treatment in genetically related diseases like neurodegenerative disorders and cancer (Hulka 1990). This shift has paved the way to the field of personalized medicine. Also biomarkers role in predicting toxic responses, diagnosing a disease, screening, risk assessment, staging, grading of tumors or cancers, clinical diagnostics, monitoring therapy, monitoring recurrent diseases, early detection of disease, patient selection, and stratification, predicators of disease, predicting the genetically related diseases, and predicting efficacies of both drugs and vaccines revolutionized biopharmaceutical science and laboratory medicine.

3.1.1 Definition

Biological markers (biomarkers) are defined as "cellular, biochemical or molecular alterations that are measurable in biological media such as cells, human tissues or fluids." Most recently, the definition has been broadened to include biological characteristics that can be specifically measured and evaluated as an indicator of normal biological or pathogenic processes, or pharmacological responses to a therapeutic intervention (Naylor 2003).

3.1.2 Evolution of Biomarkers

Biomarkers are either used for the diagnostic or predictive purpose. Testing of biological fluids for diagnostic and predictive purposes started 6000 years ago (Armstrong 2007). Documented evidence indicates the use of body fluids like urine and liver for diagnosing diseases by Sumerian, Babylonian, Egyptian, and Hindu physicians. Sumerian and Babylonian physicians used body fluids other than physical evidence of disease to make a clinical decision. Physicians would let the patient breathe into sheep's nose, later slaughter an animal, remove the liver and carefully inspect evidence of diseases to diagnose patients and subsequent outcomes for treatment. Physicians of Egypt Pharaohs used wheat and barley seeds to test hormones in the body fluids. Patients' urine was added to a bag containing wheat and barley seeds, if seeds germinated woman was pregnant. If barley seeds germinated first it can be a male fetus, but if wheat seeds germinate first it indicates a female fetus. The ability to attract black ants towards urine containing sugar was used as a diagnostic test for diabetes mellitus (Armstrong 2007; Winsten 1969). The chapter provides further details on types of biomarkers, the volume of biomarkers business, stages in development of biomarkers, and applications of biomarkers.

3.2 Types of Biomarkers

Biomarkers were classified according to usage by various groups like Abdul Baset Halim (2011), Frank and Hargreaves (2003), Kumar and Sarin (2009), and Turner and Hellerstein (2005) (Fig. 3.1). Abdul Baset Halim (2011) classified biomarkers as safety and efficacy; Kumar and Sarin (2009) classified biomarkers as DNA, RNA, and protein; Frank and Hargreaves (2003) classified biomarkers as clinical, microimaging, and next generation; and Turner and Hellerstein (2005) classified biomarkers as static and kinetic biomarkers.

3.2.1 Classification of Biomarkers by Kumar and Sarin (2009)

Differential expressions of the DNA, RNA, and proteins in normal and diseased cells are taken into consideration for the classification of biomarkers. The different types of biomarkers according to Kumar and Sarin (2009) are DNA, RNA, and protein.

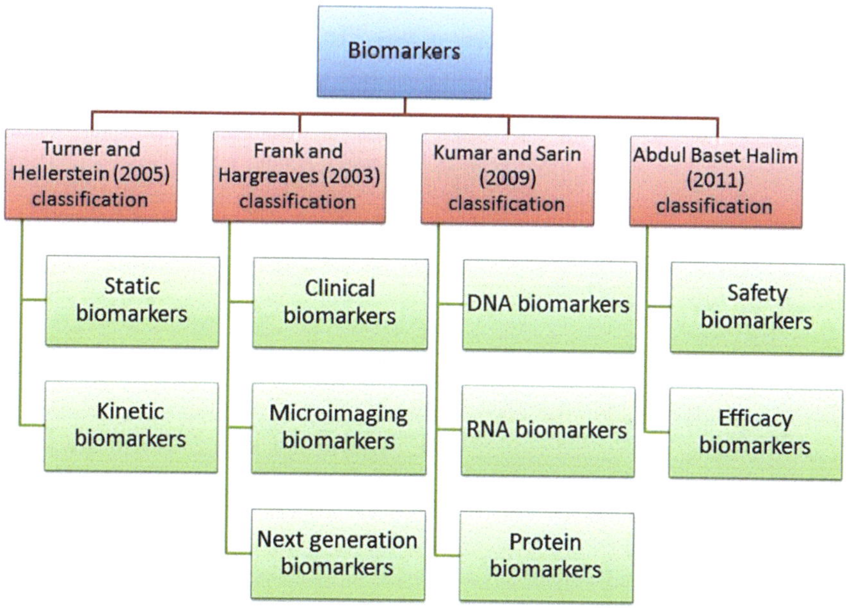

Fig. 3.1 Biomarkers and their classification

3.2.1.1 DNA Biomarkers

Mutated tumor-suppressor genes; oncogenes; mismatch repair genes; oncoviral markers; SNPs associated genes of colon, lungs, head, neck, bladder, etc. related to the genome and mitochondrial DNA are known as DNA biomarkers. Increased DNA levels of the above genes in serum are associated with various types of cancers, autoimmune diseases, and sepsis.

3.2.1.2 RNA Biomarkers

High-throughput technologies were used to screen the mRNA molecules based on the comprehensive mRNA expression which are acting as the indicators for toxicity and efficacy of drugs. These mRNA molecules are called RNA biomarkers. Most of the RNA molecules are having their application in the field of cancer. Many of these RNA-based biomarkers in the form of multi-gene molecular patterns or "fingerprints" (Gray and Collins 2000) are undergoing clinical evaluation.

3.2.1.3 Protein Biomarkers

Protein based on fingerprints from 2D –PAGE, MudPIT, reverse-phase microarray, SELDI-TOF, immuno-PCR, field-effect transistor, phosphorylation-dependent signaling cascades, quantum dots identify differentially expressed proteins characteristics of normal and diseased cells. These are known as protein biomarkers.

3.2.2 Classification of Biomarkers by Abdul Baset Halim (2009)

Abdul Baset Halim used safety and efficacy as the criteria to classify biomarkers. There are two types of biomarkers—safety and efficacy.

3.2.2.1 Safety Biomarkers

Specialized function tests are applied in different therapeutic areas to detect unique safety toxicities. Markers identified with these unique safety toxicities can be applied as biomarkers to identify toxicity as early as possible in clinical development. This information can be used to constantly monitor the safety of common vital organs. Safety testing generates the following biomarkers—liver safety biomarkers, renal safety biomarkers, hematology safety biomarkers, bone safety biomarkers, metabolic safety biomarkers, and specific safety biomarkers (Table 3.1).

Table 3.1 Different types of biomarkers which were used in clinical trials during drug development program to measure safety and efficacy

S. No.	Class	Sub-class	Biomarkers examples	
1	Safety biomarkers	Liver safety biomarkers	• Alanine transaminase (ALT) • Serum aspartate transaminase (AST) • Alkaline phosphatase (ALP)	• Gamma-glutamyl transferase (GGT) • Bilirubin
		Renal safety biomarkers	• Blood urea nitrogen (BUN) • Serum creatinine • Glomerular filtration rate (GFR) • Creatinine clearance • Serum electrolytes—sodium, potassium, chloride, and bicarbonate • Complete urine analysis—color, pH, specific gravity, glucose, proteins, ketone bodies, and microscopic examination for blood, leukocytes, casts • Cystatin c	• β 2-microglobulin • Uric acid • Clusterin • N-acetyl-beta-D-glucosaminidase • Neutrophil gelatinase-associated lipocalin (NGAL) • N-acetyl-β-D-glucosaminidase (NAG) • Kidney injury molecule-1 (KIM-1)
		Hematology safety biomarkers	• Complete blood count • Total hemoglobin • Hematocrit • Red cell count • Mean red cell volume • Mean cell hemoglobin • Red cell distribution width%	• Mean cell hemoglobin concentration • Total white cell count • Differential white cell count—neutrophils, lymphocytes, basophils, eosinophil, monocytes, and platelets
		Bone safety biomarkers	• Serum calcium	• Inorganic phosphates
		Metabolic safety biomarkers	• Blood glucose • Triglycerides(TG) • Total cholesterol	• Low-density lipoprotein cholesterol (LDLC) • High-density lipoprotein cholesterol (HDLC)
		Specific safety biomarkers	• Serum immunoglobulin levels • C-reactive protein (CRP) • Fibrinogen • Thyroid stimulating hormone (TSH) • Thyroxine • Testosterone	• Insulin • Lactate dehydrogenase (LDH) • Creatine kinase (CK) and its isoenzymes • Cardiac troponin (cTn) • Methemoglobin

(continued)

Table 3.1 (continued)

S. No.	Class	Sub-class	Biomarkers examples	
2	Efficacy biomarkers	Surrogate biomarkers	• Blood pressure (BP) for myocardial infarction • Cholesterol • LDLC • Triglycerides • Blood glucose • Glycosylated hemoglobin (HbA1c) • Arterial plaque thickness	• CD4 count or viral load for HIV response • HCV RNA viral load for HCV response • Bacterial count • Tumor size • Bone mineral density
		Predictive biomarkers	• Cytochrome P450-2D6 (CYP2D6) • LDLC	• HbA1c • CYP2C19
		Pharmacodynamic (PD)	• Pharmacokinetic (PK) measurements	
		Non-imaging biomarkers	• Proteins • Cytokines • Enzyme activity in serum • CSF or tissue lysates	• Proteins by immunohistochemistry (IHC) • DNA and RNA gene expression
		Prognostic biomarkers	• HER2/neu • c-KIT	• EGFR1

3.2.2.2 Efficacy Biomarkers

Efficacy means-testing the benefit or harm of a therapeutic agent, classifying populations as responders and non-responders, and explaining the mode of action of the drug, predicting the outcome. This information or biomarkers can be used to estimate the efficiency of therapeutic agents.

3.2.3 Classification of Biomarkers by Frank and Hargreaves (2003)

Biomarkers are valuable and help to prioritize drug discovery using various methodologies and technologies. This information can be used to anticipate, plan, and substantiate drug discovery. This group classified biomarkers as clinical, microimaging, and next generation.

3.2.3.1 Clinical Biomarkers

Markers that reflect the relevant changes in terms of the clinical outcome when a drug is used in the drug development process are known as clinical biomarkers. There are different types of clinical biomarkers Natural history or Type 0 biomarkers, Drug Safety or Type I biomarkers, Surrogate or Type II biomarkers (Table 3.2).

3.2.3.2 Microimaging Biomarkers

Images generated in vivo assessment which are reflecting changes in terms of clinical outcome, when a drug is used in drug development program using micro-CT, micro-ultrasound, and micro-PET detectors and reduce decision-making are known as microimaging biomarkers. The different types of microimaging markers are cardiology, neurology, oncology, psychiatry, depression and pain, osteoarthritis biomarkers (Table 3.2).

3.2.3.3 Next-Generation Biomarkers

Markers like pharmacogenetic/pharmacogenomic, genotypic drug metabolism and transport, haplotype and SNP, RNA expression profiling, metabolomics, and proteomics that are generated/identified using next-generation technologies for the clinical outcomes during the development program are called next-generation biomarkers. The different types of biomarkers are pharmacogenetic/pharmacogenomic; genotypic drug metabolism and transport; haplotype and SNP; RNA expression profiles; and metabolomics and proteomics (Table 3.2).

3.2.4 Classification of Biomarkers by Turner and Hellerstein (2005)

Present tools were able to advance little in drug discovery and development programs. There is a requirement to measure the flux in intact living systems to understand the precise volume or content. There are two types of biomarkers, namely static and kinetic.

3.2.4.1 Static Biomarkers

Markers that track static measures in a living system are known as static biomarkers.

Table 3.2 Biomarkers which anticipate (clinical biomarkers), reflect (microimaging biomarkers), generate (next-generation biomarkers) clinical outcome during drug discovery program

S. No.	Class	Sub-class	Biomarkers examples	
1	Clinical biomarkers	Type 0 biomarkers	• Metallopeptidase inhibitor 1 (TIMP1)	• EGFR1
		Type 2 biomarkers	• Refer Table 3.1 safety biomarkers section	
		Type 3 biomarkers	• Refer Table 3.1 surrogate biomarkers section	
2	Microimaging biomarkers	Cardiology biomarkers	• Ultrasonography • Magnetic resonance imaging(MRI) • Thermography • Fluorodeoxyglu cose positron emission tomography (FDG-PET) • Intravascular ultrasonography (IVUS)	• Conventional coronary angiography (QCA) • Computerized tomography • C-reactive peptide • QT prolongation • Troponin T
		Neurology biomarkers	• Magnetic resonance spectroscopy (MRS) • Single photon emission computed tomography (SPECT)	• High-resolution 3 tesla (3T) MRI • Diffusion and perfusion MRI • Perfusion CT
		Oncology biomarkers	• Multi-dimensional imaging • Multi-modal imaging • High-resolution CT	• FDG–PET • Fluorine Labeled thymidine (FLT)
		Psychiatry biomarkers	• Serum cortisol • Adrenocorticotropic hormone • Clonidine stimulated growth hormone	• Neuro transmitter mapping • Functional mapping (functional MRI)
		Depression and pain biomarkers	• FMRI picture stimuli	• Functional PET
		Osteoarthritis biomarkers	• Osteoarthritis imaging initiative • High field strength 3 Tesla • Contrast MRI sequence (dGEMERIC)	• Chelated gadolinium (Gd-DTPA, gadolinium diethylene-triamine-penta-acetic acid)
		Microimaging biomarkers	• Translational research technologies • Small animal micro-CT	• Micro-ultra sound • Micro-PET detectors

(continued)

Table 3.2 (continued)

S. No.	Class	Sub-class	Biomarkers examples	
3	Next-generation biomarkers	Pharmacogenetic/ pharmacogenomic biomarkers	• Genotypes • Single nucleotide polymorphisms • Haplotypes • Cytochrome P450 enzyme CYP2D6 • Flavin containing monooxygenase-3 (FMo3) or CYP1A2	• Haplotype and SNP biomarkers–β_{2}_AR • RNA expression profiling–Quantitative trait loci (QTL's)
		Metabonomics and proteomics biomarkers	• Laser scanning cytometry • Liquid chromatography-mass spectrometry (LC-MS)	• NMR
		Environmental factors biomarkers	• Single ICD-9CM code for the dysmetabolic syndrome	• Diabetes-environmental impact of diet and exercise

3.2.4.2 Kinetic Biomarkers

Markers that measure the kinetics of the intact system to track the flux preceding changes in pool size or content, i.e., precise physical-chemical measurement are known as kinetic biomarkers.

3.3 Volume of Business with Biomarkers in Medicine and Pharmaceutical Industry

Adrian Dawkes, a consultant at PharmaVenture has given the strategic importance of biomarkers to the pharmaceutical industry (Dawkes 2007). The use of biomarkers in drug development process maximized safety and efficacy for the patient, generating revenues and margins to companies. By the introduction of biomarkers in the drug development process from the year 2001–2007, the average revenue generated for each drug was nearly $ I billion (Table 3.3). In turn, the introduction of biomarkers, at which every phase and "Go" and "No Go" decisions save nearly $ 100 million per drug. Based on the margins that are generated by biomarkers as products these can be classified as blockbusters and niche busters.

Table 3.3 Volume of business generated by some important drugs due to the introduction of biomarkers as a companion in a drug development program

S. no.	Drug	Revenue	Year
1	Herceptin®	$ 4 billion	2007
2	Gleevac®	$ 2.5 billion	2006
3	Tamofexin	$ 630 million	2001

3.3.1 *Blockbusters*

Products that are protected as patents and allow pharmaceutical companies to achieve revenues on the original R&D investments, permitting new investment into the next generation of therapeutics are known as blockbusters.

3.3.2 *Niche Busters*

Products that are developed only for positive responders in a patient population, whereby drugs are prescribed and funded only for those patients, such products are known as niche busters. Thereby potential market/revenue of niche busters dependent on a number of patients leading to personalized medicine age.

3.4 Intellectual Property on Biomarkers

Academic research institutes are discovering and filing intellectual properties in the biomarkers area. Intellectual property on biomarkers offers commercial opportunities and pharmaceutical companies are taping these rich biomarkers for the development of therapeutics. IP filings were reported to be in fields of cancer, inflammation, sepsis, Alzheimer's, cardiovascular, diabetes, neurological, renal, autoimmune, rheumatoid arthritis (RH), HIV, stroke, leukemia, etc.; among them cancer showed more than 50% IP filings. Among the fields of cancer, breast and prostate cancers showed more than 20% IP filings, ovarian and bladder showed more than 10%, pancreatic and liver showed more than 5%. The rest of the fields like lung, colon, endocrine, cervical, colorectal, nasopharyngeal, head and neck, testicular showed less than 5% IP filings (Dawkes 2007).

3.5 Stages in Development of Biomarkers

The development of biomarker for normal biological or pathogenic processes, or co-development of a companion biomarker for a drug development program is essential. The task of identifying biomarkers involves three stages: discovery, assay development, and commercialization (Fig. 3.2).

3.5.1 *Discovery*

The discovery stage includes identification of biomarkers, Study Protocol Proposal "fit for purpose." Identification of biomarker includes discovering or identifying biomarkers, "Study Protocol Proposal" is used for method development in animal models or cell cultures, and "fit for purpose" is used to link biomarker with biology and clinical endpoints.

3.5.1.1 Identification of Biomarkers

Biomarkers are discovered/identified by DNA microarray, real-time PCR; 2D-gel electrophoresis and mass spectrometry (MS) to identify differentially expressed markers in normal and diseased cells. These differentially expressed markers are subjected to digestion in gel, or western blot or immunohistochemical analyses. This is followed by "Study Protocol Proposal" for method development and "fit for purpose."

3.5.1.2 Study Protocol Proposal

The goal of the proposal, is to define an analytical assay. The proposal is prepared to study the protocol and validate the target identified (Cummings et al. 2010). The following are key sections of the proposal:

- Identification of samples in which the biomarkers will be measured.
- Statistical analysis plan, outlining method and acceptance criteria for biomarkers.

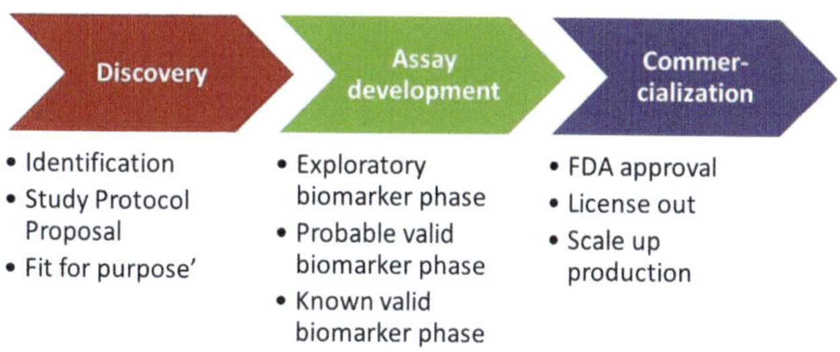

Fig. 3.2 Stages in the development of biomarkers

- Define data set for validation of biomarkers.
- The rationale for sampling and, exposure data related to biomarkers.
- Define the use of biomarker as a mechanistic, diagnostic, or predictive candidate.
- Power calculations for accuracy, specificity, and sensitivity of biomarkers.
- Plan for future cross-species comparison of biomarkers.
- Preliminary dose-ranging step for biomarkers.
- Gain the knowledge about compound pharmacology used in testing biomarkers.
- Gain the knowledge about compound toxicity used in tesing biomarkers.
- Number of compounds tested for biomarkers.
- Replicate animals in the final study on biomarkers.
- Necropsy plan, clinical pathology, and histopathology studies for biomarkers.

3.5.1.3 Fit for Purpose

"Fit for purpose" approach has emerged to guide biomarker method validation. Ideally, validation progress down two parallel tracks one experimental and other operational. Cummings et al. (2010) stated that "… the first is to establish the purpose of the method and agree upon outcomes, target values or acceptance limits; whereas the second is to characterize the performance of the assay by experimentation. The critical step is to evaluate the technical performance against the predefined purpose. If the assay can define to expectations then it is deemed to fit for that purpose, if not, then it cannot be fit for the specific purpose…".

There are five stages in "fit for purpose" (Lee et al. 2006, 2007).

Stage 1: It defines purpose and selection of the candidate assay.

Stage 2: Appropriate reagents and components are assembled to write the method validation plan and decide upon the final classification of the assay.

Stage 3: Evaluation of "fit for purpose" via the performance of the experimental phase culminates in writing a standard operating procedure.

Stage 4: Robust assays were carried out in the clinical context to identify patient sampling issues, such as collection, storage, and stability.

Stage 5: Assay enters testing and batch-to-batch QC issues which can be fully explored.

The different types of fit for purpose assays are quantitative, definitive quantitative, relative quantitative, and quasi-quantitative assays. A definitive quantitative assay uses calibrators and a regression model to calculate absolute quantitative values for the unknown with the reference, which is a fully characterized standard and which are representatives of biomarkers. A relative quantitative assay uses a response concentration calibration with reference standards that are not fully representative of the biomarker. A quasi-quantitative assay does not employ a calibration standard but has a continuous response that can be expressed in terms of a characteristic of the test sample. Quantitative (categorical) assays can either be described as ordinal reliant on discrete scoring scales like those used in immunohistochemistry

(IHC) or nominal that pertains to a yes/no situation; for example, the presence or absence of a gene product (Lee et al. 2006, 2007).

3.5.2 Assay Development

Exploratory biomarker phase, probable valid biomarker phase, known valid biomarker phase are the three phases of the assay development stage. In the exploratory biomarker phase, a feasibility assay is developed and "A Go" and "No Go" decision is taken for biomarker using nonclinical samples. In the probable valid biomarker phase, feasibility assay is refined further with limited clinical samples to establish clear performances as a preclinical assay to track the efficacy. This information is used to complete the validation report, standard operating procedures (SOPs) for planning, implementing, and employing biomarkers for the assay. The known valid biomarker phase, is where FDA reviews biomarker and then the biomarker is known as "known valid biomarker". The assay developed at this phase is used for testing clinical samples. Robust assay with stringent requirements on accuracy, sensitivity, specificity, precision, reproducibility (inter- and intra-lab evaluation) are performed for the validity of biomarkers. Data is gathered; a validation report and an SOP are prepared and submitted for approval at FDA. Center for Medicare and Medical Services (CMS) under the Clinical Laboratory Improvement Amendment (CLIA) Rules review for necessary improvement.

3.5.3 Commercialization of Biomarker

After FDA approval, biomarker developers either out a license or seek a commercial partner to scale up production. Later launch approved in vitro diagnostic (IVD) or analyte-specific reagent (ASR) or kit product of biomarker for robust assay in clinics, hospitals, reference laboratories for diagnostic and therapeutic purposes. A comprehensive collection of clinical, preclinical, and exploratory markers pooled into a database called GVK Bio Online Biomarker Database (GOBIOM) associated with different therapeutic areas. This database is helping the practitioners to visualize the effects of those markers.

3.6 Applications

As in all fields of science, the use of biomarkers is rapidly increasing; the reliability and success of a biomarker development program depend on the applications (Fig. 3.3).

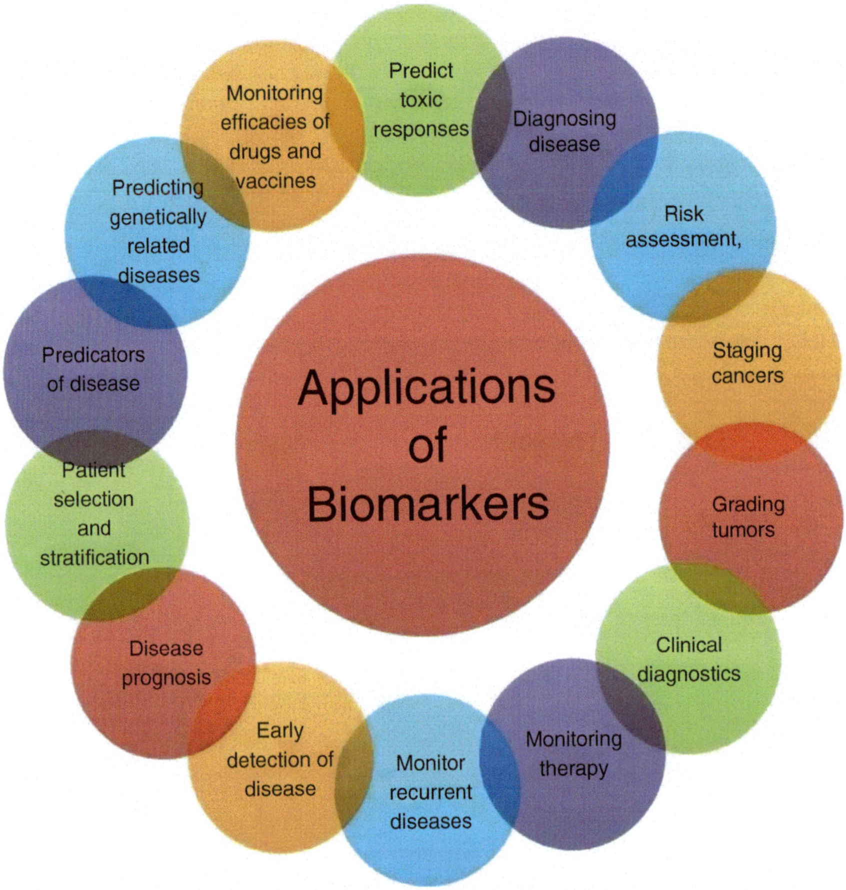

Fig. 3.3 Application of biomarkers

1. Early cellular responses can be studied using in vitro systems and biomarker, that may help to predict toxic responses in vivo.
2. Biomarkers are useful in the field of medicine for diagnosing a disease that is either produced by the diseased organ (e.g., tumor) or by the body in response to that particular disease to understand the whole spectrum of the disease process. During diagnosis, markers could be used for screening, risk assessment, staging, and grading.
3. Biomarkers have been widely used in clinical diagnostics.
4. Biomarkers are used for the selection of initial therapy, monitoring therapy, selection of additional therapies, and to monitor recurrent diseases (Naylor 2003).
5. Biomarkers are used to identify subgroups of patients who respond to therapies and interventions in different ways.

6. Biomarkers aid in the early detection of disease and in the investigation of therapies/interventions aimed at reducing the risk of disease.
7. Biomarkers are used in the field of medicine for disease prognosis, besides as metabolism biomarkers. Other areas where biomarkers are used is in patient selection and stratification (e.g., CCR5 receptor).
8. Biomarkers are much better predictors of disease (illness) and death than self-reported health status.
9. Genomic biomarkers are extensively used in predicting genetically related diseases.
10. Biomarkers are increasingly being used by researchers associated with industry, universities, and government for studies and have proven to be cost-effective.
11. Biomarkers are reliable for monitoring, developing, and predicting efficacies for both drugs and vaccines.
12. Biomarker immunoassays are revolutionizing biopharmaceutical science and laboratory medicine. The antibody/antigen-based assays have made possible to measure minute amounts of proteins in a diverse array of biologically relevant samples. The sensitivity and specificity of traditional immunoassays are frequently limited by endogenous and exogenous interferences that cannot be eliminated and researchers struggle to identify and eliminate their effects.

3.7 Future Perspectives of Biomarkers

The future of biomarkers appears to be a mixture of excitement and uncertainty. This ambiguity is due to many disciplines, the practitioners and scientists contribute to identify novel biomarkers. Disciplines may include medicine, pharmaceutical industry, diagnostics, ecotoxicology, environmental monitoring, and environmental exploration. To provide direction, clarity, to achieve the goals; diverse group of people with sets of skills, a variety of tools are needed for a better contribution. Advanced scientific technologies and changes in currently used methods will play an efficient role in identifying novel biomarkers.

Future applications of biomarkers in medicine and pharmaceutical industry is to identify the critical illness using specific biomarkers that identify pathophysiologic effects for a particular disease and to provide selective and guided therapy. There is a need for several initiatives to consider building an excellent surrogate endpoint for biomarker development. Consideration of consortia formation is another aspect to be discussed, particularly concerning biomarker databases, nomenclature, and data visualization.

Few biomarkers of environmental monitoring can provide an early warning of deleterious effects on biological systems and for estimating those biological effects due to contaminants. Ecotoxicology biomarkers can create awareness about the toxic effects caused by the natural or synthetic pollutants which constitute the contamination of ecosystems, animals, plants human, and microbial populations. Some of the environmental exploratory biomarkers mostly hydrocarbons may be used in petroleum exploration.

3.8 Conclusions

Early detection of cancer and the drug development process requires biomarkers. Discovery, assay development, and commercialization are the three stages of biomarker development. Biomarkers can be identified, analyzed, developed by "Study Protocol Proposal" and "Fit for Purpose" in the discovery stage. Exploratory, probable valid, and known valid biomarker phases are the three phases used to validate biomarkers in the assay development stage. Finally, commercialization is the last stage where biomarkers are approved by FDA and CMS according to CLIA rules and used clinically for diagnostic and therapeutic purposes.

Acknowledgments NNRR, CS, and STPRCPK are grateful to GITAM (Deemed-to-be-University) for providing necessary facilities to carry out the research work and for extending constant support.

Conflict of Interest The authors declare that there is no potential conflict of interest.

References

Abdel-Baset H (2011) Biomarkers in drug development: a useful tool but discrepant results may have a major impact. In: Kapetanovic I (ed) Drug discovery development—Present and future, 1st edn. In Tech, Rijeka, pp 401–424

Armstrong JA (2007) Urinalysis in Western culture: a brief history. Kidney Int 71:384–387

Cummings J, Raynaud F, Jones L, Sugar R, Dive C (2010) Fit-for-purpose biomarker method validation for application in clinical trials of anticancer drugs. Br J Cancer 103:1313–1317

Dawkes A (2007) The strategic importance of biomarkers to the pharmaceutical industry. PharmaVentures. http://files.pharmaventures.com/biomarkers_pharmaventures.pdf

Frank R, Hargreaves R (2003) Clinical biomarkers in drug discovery and development. Nat Rev Drug Discov 2(7):566–580

Gray JW, Collins C (2000) Genome changes and genetic expression in human solid tumours. Carcinogenesis 21:443–452

Hulka BS (1990) Overview of biological markers. In: Hulka BS, Griffith JD, Wilcosky TC (eds) Biological markers in epidemiology, 1st edn. Oxford University Press, New York, pp 3–15

Kumar M, Sarin SK (2009) Biomarkers of diseases in medicine. Curr Trends Tech Sci, 403–417

Lee JW, Devanarayan V, Barrett YC, Weiner R, Allinson J, Fountain S, Keller S, Weinryb I, Green M, Duan L, Rogers JA, Millham R, O'Brien PJ, Sailstad J, Khan M, Ray C, Wagner JA (2006) Fit-for-purpose method development and validation for successful biomarker measurement. Pharm Res 23(2):312–328

Lee JW, Figeys D, Vasilescu J (2007) Biomarker assay translation from discovery to clinical studies in cancer drug development: quantification of emerging protein biomarkers. Adv Cancer Res 96:269–298

Naylor S (2003) Biomarkers: current perspectives and future prospects. Expert Rev Mol Diagn 3:525–529

Turner SM, Hellerstein MK (2005) Emerging applications of kinetic biomarkers in preclinical and clinical drug development. Curr Opin Drug Discov Dev 8:115–126

Winsten S (1969) The skeptical chemist. Clin Chem 15:737–744

Chapter 4
Prognostic Molecular Markers for Gastrointestinal Cancer

Achanta Jagadeesh, G. Mohana Sheela, B. Pratap Naidu, and Pallaval Veera Bramhachari

Abstract Cancer is considered as the most dreadful diseases worldwide. The rate of mortality is increasing every year globally. Among the various cancers, gastric cancer is the fifth most common cancer-causing after the various other cancers like lung, breast, prostate, and even the abdomen. This cancer is the third most cause of cancer death. Various environmental factors like smoking, the role of diet, and some bacterial infections result in gastric cancer. The gastric cancer is unrecognizable at its early stages and is diagnosed only at the advanced stages where the risk of saving a person is unimaginable. There are certain genes in which codes for gastric cancer are mutated which leads to gastric cancer. Identifying correct genes through the biomarkers will help in eradicating the disease at its early stage. Transforming this information from patient care to diagnostic tools remains a challenge for many researchers. Researchers are currently working to translate molecular information into the development of drugs. Before identifying the correct drug, researchers need to focus on identifying the genes which cause gastric cancer and even the pathways associated with detecting cancer at an early stage. Current generation researchers are working on next-generation sequencing which has led to molecular classification systems that are used in designing new targeted therapies and are implemented in clinical trials. This chapter will focus on the latest applications/techniques required in identifying various molecular markers for gastric cancer and even certain metabolic pathways/signalling pathways will be identified/reviewed to identify the correct diagnosis for gastric cancer. This chapter will even focus on the biomarker-targeted therapies that are involved in the treatment of gastric cancer.

Keywords Gastric cancer · Diagnostic markers · Signalling pathways · Enzymes · Pathology · Disease

---

A. Jagadeesh
Tumor Microenvironment Global Core Research Center, College of Pharmacy, Seoul National University, Seoul, South Korea

G. M. Sheela · B. P. Naidu · P. Veera Bramhachari (✉)
Department of Biotechnology, Krishna University, Machilipatnam, Andhra Pradesh, India

P. Veera Bramhachari, N. R. R. Neelapu (eds.), *Recent Advancements in Biomarkers and Early Detection of Gastrointestinal Cancers*, Diagnostics and Therapeutic Advances in GI Malignancies, https://doi.org/10.1007/978-981-15-4431-6_4

4.1 Introduction

Gastric cancer is one of the fatal diseases in the world; it is the second largest cancer in cancer deaths (Jemal et al. 2011). According to WHO, it is reported as 24,590 are affected; nearly 10,720 GC deaths are diagnosed in the USA. Gastric cancer is a nonspecific symptomatic disease; it provides a potential platform to transform and attain oncogenicity (Wagner et al. 2010). The normal symptoms are like stomach ache, anorexia, weight loss, and difficulty in ingestion. The major causes of gastric cancer are diet, *Helicobacter pylori* infection, atrophic gastritis, and intestinal metaplasia. Four somatic modifications in gastric cancer are observed like EBV, microsatellite instability (MSI), genomic stability, and chromosomal instability (CIS). The occurrence of GC was categorized into intestinal and diffuse types by Lauren classification (Wagner et al. 2010). The intestinal GC is glandular with variable differentiation and usually observed in old patients which are formed as a result of causative effects in GC. The diffuse gastric carcinoma, usually motile neoplasm, is infiltrated to different locations of gastric walls. It was observed in young patients.

Diagnosis plays an important role in disease progression and prevention. The conventional methods are used to diagnose GC by laparoscopy and gastroscopy. The reoccurrence of GC was diagnosed by CT scan, echoendoscope (Mihaljevic et al. 2013). The major drawback in diagnosing GC cancer was identifying the stages and reoccurrence of GC. The interventions are measured by various biomarkers and quantified through the pharmacological or normal biological responses. There are other biomarkers like DNA, exosomes, noncoding RNAs, etc. As a result, proteins and genes are exploited to diagnose GC. The biomarkers are characterized into four types such as diagnostic, prognostic, predictive, and therapeutic biomarkers (Matsuoka and Yashiro 2018).

4.2 Conventional Prognostic Markers

The conventional prognostic or biomarkers are identified and characterized based on the surface proteins and genes. There are different cell types that favor the disease progression and pathogenesis (Lin et al. 2012). As a result, they are identified as potential prognostic markers. Likewise, metastatic genes-signalling mediators, immune checkpoint, microsatellite instability.

4.3 Unconventional (Noninvasive Prognostic Markers)

The noninvasive prognostic markers are widely identified by the body fluids like blood, urine, and other body fluids. These are characterized by the site of origin rather than its biological significance.

The major markers are characterized like CTCs, circulating cell-free DNA, miRNA, long-noncoding RNA, and exosomes. These are used as liquid biopsy and help in the identification of stages and quantify the gastric cancer (Siravegna et al. 2017).

4.3.1 Metastatic Genic Prognostic Marker

The metastatic genes, which initiates the transformation of oncogenes with various RTKs and signalling mediators. Thus, the mediators are acting as biomarkers for predictive, prognostic, and diagnostic markers.

4.3.1.1 HER2

HER2 is one of the RTKs, the overexpression potentiates the transformation of oncogene by activating signalling cascade. It is the first biomarker found in GC with poor prognosis and the study highlights the HER amplification is found in patients, who are ranging from 6 to 23%. The HER2 located in the gastroesophageal junction compared to the distal end. The HER2 overexpression is a result of mutations in the erB2 gene that leads to early-stage carcinogenesis.

The role of a biomarker is a bit controversial, in spite of that it has poor prognostic value, it is measured by a chemotherapeutic drug called lapatinib and trastuzumab. The drugs inhibit the HER2 and as a result, progression-free survival was enhanced. The HER2 overexpression was inhabited by various other drugs, and hence, the HER2 acts as a target to the drugs and inhibits the overexpression and inhibits the carcinogenesis (Gomez-Martín et al. 2014). As a result, it can act as a potential prognostic biomarker.

4.3.1.2 MET

MET, is one the receptor tyrosine kinases identified as hepatic growth factor (HGF), it activates various signalling cascade. As a result, it leads to cell proliferation and cell growth. The overexpression of MET leads to over proliferation, angiogenesis, and migration; hence, it is responsible for the poor prognosis of GC (Matsumoto et al. 2017). The MET is characterized as a prognostic and predictive marker for GC by activating signalling cascade likewise, HGF/c-Met signalling cascade. The high serum HGF can be a possible prognostic marker, where the low levels of HGF are treated with chemotherapeutic drug, i.e., trastuzumab with positive outcomes. These results can highlight the significance of MET as a potential prognostic marker.

4.3.1.3 VEGF

Vascular endothelial growth factor is one of the growth factor responsible for angiogenesis. The neovascularization provides a platform for the formation of new blood vessels for the tissues to attain a normal physiological state. In GC, the VEGF promotes tumor proliferation, survival, and migration by the various signalling cascades. The VEGF has different isoforms, in the recent study highlights the VEGF-2 has potential prognostic value in a ramucirumab treatment. The VEGF-D also can be a promising prognostic marker in the ramucirumab-treated patients (Matsuoka and Yashiro 2018).

4.3.2 MSI

The short repeated nucleotide sequences around 1–6 which are located in the noncoding and protein-coding sequences regulate the expression by addition or deletion of repeating units. As a result, it leads to genomic instability and tumorigenesis. The Gastrointestinal Cancer incidence is estimated by the high and low MSI; the low MSI was characterized by less than 30%; and the high MSI was characterized by more than 30% (Pinto et al. 2000). In Gastrointestinal Cancer, the epigenetic silencing of MLH1 by hypermethylating its promoter. The MSI regulates silencing and activating various expressions of targets which mediates the GC carcinogenesis. The mutations in PIK3CA are observed in MSI-positive GC, which highlights the genomic instability and regulates the targets and their expression. This can be a prognostic biomarker due the differential expressions in the high and low MSI conditions and the extent of chemotherapy treatments and can obtain better clinical outcomes (Smyth et al. 2017).

4.3.3 Genetic Polymorphism

The genetic polymorphism in the carcinogenesis plays an important role in GC; the genetic polymorphism is mainly characterized as SNPs. In GC, the functional similarity between IL-beta and IL-RN in the *Helicobacter pylori* infection induces the progression of chronic gastritis and GC in the Algerian population. CD44, a glycoprotein highly expressed in the GC, has different isoforms involved in GC (Suenaga et al. 2015). In GC, CD44 SNP rs187116 assumed to have high expression and it can be a prognostic biomarker. Apart from above, other SNPs like TP53, CDH1, and ARID1A are putative targets for prognostic biomarkers as SNPs.

4.3.4 Long-Noncoding RNA

The noncoding RNA containing more than 200 nucleotides are termed as long-noncoding RNA. lncRNA has diverse functions and it regulates transcription, splicing, chromatin remodelling, and post-translational modification. It acts as an oncogene and tumor suppressor. 135 lncRNA are found in dysregulated GC (Fang et al. 2015). As a result, it leads to tumorigenesis, metastasis, and prognosis. The minimal expression of lncRNA like AI364715, GACAT1, and GACAT2 in GC acts as a prognostic biomarker.

4.3.5 Immune Checkpoint

The immune cell inhibition plays a key role in tumor progression and also in GC. The immune activation is attained by PD-1 and PD-2 molecules on the T and B cell surfaces (Sharpe et al. 2007). But in the cancer progression, the T cell and B cell activation are inhibited by PD-L1 and PD-L2. As a result, the activation of cytotoxic T cells are inhibited and immune resistance towards tumor facilitates the tumor survival and progression (Gu et al. 2017). In GC, the inhibition of immune activation in the mucosa of the gastrointestinal tract has poor disease prognosis. PD-L1 is expressed in more than 40% in EB positive condition. According to the study, the PD-L1 expression is high in MSI high condition. The patients treated with the PD-1 inhibitor pembrolizumab have a better survival rate with untreated patients. Therefore, PD-1 can be a potential prognostic biomarker for GC.

4.4 Noninvasive Biomarkers

The differentiation of solid tumors from the patient sample is very tough and determining the stage of the cancer is challenging. To overcome the limitation, researchers found liquid biopsy to characterize and identify the tumor concerning stages and progression. For the liquid biopsy, blood and other body fluids are used.

4.5 CTCs

Circulating tumor cells (CTCs) are the single cells or clusters, identified in the bloodstream which is disseminated from the tumor cells. The CTC can be found in all stages of cancer, majorly in neoplasms. It has metastatic and stems like properties; facilitate the tumor metastasis and tumor renewal respectively. In Gastrointestinal Cancer, the CTCs assumed to have CD44 and other EMT markers which can be evident to have stemness and metastatic conditions.

4.6 Circulating Cell-Free DNA (cfDNA)

The blood is a major carrier of different kinds of cells from normal and cancer cells. The cfDNA is characterized as cell-free extracellular DNA. It was released from neoplasms, primary tumors, and metastatic tumors. The main advantages are specificity, limited sample volume. In GC, the cfDNA originated from the methylated promoter regions of cells and identified by the PCR technique. Likewise, a specific region APC1 in serum and RASSF1A promoter methylations are usual epigenetic modifications of cfDNA. Surprisingly, the study finds the presence of EBV DNA in cfDNA of GC. This infers the potentiality of cfDNA in the GC diagnosis and identification with high specificity.

4.7 miRNA

The short noncoding RNA consists of 18–30 nucleotides in length which bind to 3′UTR of the target sequence and regulates its translation. The miRNA is key molecule that regulates tumor activation and tumor suppression. It affects cell proliferation, cell differentiation, and cell migration. Additionally, the miRNA possesses oncogene activity with the following prognostic markers viz. miR-21, miR-23a, miR-27a, miR-106b-25, miR-199a, miR-215, miR-222-221, and miR-370 respectively. However, the miRNA possessing the tumor-suppressive activity are listed as follows; miR-29a, miR-101, miR-125a, miR-129, miR-148b, miR-181c, miR-212, miR-218, miR-335, miR-375, miR-449, miR-486, miR-512. Therefore, miRNA can be considered as a potential prognostic marker, which has different subsets and localized in blood and plasma.

In a recent study, cfmiRNA was discovered which enhances the functions of the miRNA and can be a potential prognostic marker. The cfmiRNA is derived from the tumor and secreted into the blood and circulated into body fluids. The miRNA expression profiling is examined and several miRNA are found and characterized as important prognostic biomarkers. The miRNA like miR-20b, miR-125a, miR-137, miR-141, miR-146a, miR-196a, miR-206, miR-218, miR-486-5p, and miR-506. The serum samples are analyzed by RT-PCR and NGS, which can be an ideal diagnostic marker.

4.8 Conclusions and Future Perspectives

The prognostic, diagnostic, predictive biomarkers are very essential for the identification of cancer stages and cancer progression. The prognostic markers are key for diagnosing GC in the early stages. There are different prognostic markers like conventional and nonconventional which are hugely differed based on the detections of tumor markers. The conventional prognostic markers are mainly detected by the tissue sample; unconventional prognostic markers are detected by the blood, plasma, and urine. There are few conventional markers which are potent markers like HER2, MET, VEGF-2, PD-1/2, MSI, and SNP which have a poor prognosis and high prognostic significance. However, employing miRNA, cell-free miRNAs, cell-free DNAs for the early detection of GI cancers can play important role as noninvasive prognostic markers apropos of thier specificity. As a result, the reduction in sample size makes it more prominent and unique. The stages of cancer care in the conventional biomarkers are quantified by the proteins and inhibitors like chemotherapeutic drugs. Conversely among the noninvasive biomarkers, the use cell-free DNA, and cell-free RNA, miRNA are specifically used and quantified for a specific purpose. Paradoxically, the HER2 is the early prognostic marker that does not have any clear evidence yet. The immune checkpoint inhibitors are the important prognostic markers, possessing a diversity of subsets that might play an imperative role in GI cancer prognosis, as depicted in the Table 4.1. The evolution of prognostic markers from the conventional to nonconventional is remarkable, and this highlights the potential use of the nonconventional prognostic markers.

Table 4.1 The significance of miRNA in the detection

	Types	Clinical significance	Detection	References
miRNA	miR-21, miR-23a, miR-106b-25, miR-130b, miR-199a, miR-215, miR-222-221, miR-370, miR-29a, miR-101, miR-125a, miR-129, miR-148b, miR-181c, miR-212, miR218, miR-335,miR-375,miR-449, miR-486, miR-512	Diagnostic/ prognostic	Blood/ plasma	Wu et al. (2014); Zhu et al. (2014)
cfmiRNA	• miR331, miR21	Diagnostic/ prognostic	Blood	Sierzega et al. (2017)
	• miR-20b, 125a,137, 141,146a, 196a, 206,218, 486-5p	Prognostic	Blood/ plasma	Zhang et al. (2017)
	• miR10b-5p, 132-3p,185-5p, 20a-3p,296-5p	Prognostic	Plasma	Huang et al. (2017)

Acknowledgments Achanta Jagadeesh is thankful to Tumor Microenvironment Global Core Research Center, College of Pharmacy, Seoul National University, Seoul, 08826, South Korea, and Dr. PVBC is thankful to Krishna University, Machilipatnam, for the support extended.

Conflict of Interest The authors declare that there is no potential conflict of interest.

References

Fang XY, Pan HF, Leng RX, Ye DQ (2015) Long noncoding RNAs: novel insights into gastric cancer. Cancer Lett 356:357–366. https://doi.org/10.1016/j.canlet.2014.11.005. [PMID: 25444905]

Gomez-Martín C, Lopez-Rios F, Aparicio J et al (2014) A critical review of HER2- positive gastric cancer evaluation and treatment: from trastuzumab, and beyond. Cancer Lett 351(1):30–40

Gu L, Chen M, Guo D, Zhu H, Zhang W, Pan J, Zhong X, Li X, Qian H, Wang X (2017) PD-L1, and gastric cancer prognosis: a systematic review and meta-analysis. PLoS One 12:e0182692. https://doi.org/10.1371/journal.pone.0182692. [PMID: 28796808]

Huang Z, Zhu D, Wu L, He M, Zhou X, Zhang L, Zhang H, Wang W, Zhu J, Cheng W, Chen Y, Fan Y, Qi L, Yin Y, Zhu W, Shu Y, Liu P (2017) Six serum-based miRNAs as potential diagnostic biomarkers for gastric cancer. Cancer Epidemiol Biomarkers Prev 26:188–196. https://doi.org/10.1158/1055-9965.EPI-16-0607. [PMID: 27756776]

Jemal A, Bray F, Center MM, Ferlay J, Ward E, Forman D (2011) Global cancer statistics. CA Cancer J Clin 61(2):69–90

Lin LL, Huang HC, Juan HF (2012) Discovery of biomarkers for gastric cancer: a proteomics approach. J Proteome 75(11):3081–3097

Matsumoto K, Umitsu M, De Silva DM, Roy A, Bottaro DP (2017) Hepatocyte growth factor/MET in cancer progression and biomarker discovery. Cancer Sci 108:296–307. https://doi.org/10.1111/cas.13156. [PMID: 28064454]

Matsuoka T, Yashiro M (2018) Biomarkers of gastric cancer: current topics and future perspective. World J Gastroenterol 24(26):2818–2832. https://doi.org/10.3748/wjg.v24.i26.2818. http://www.wjgnet.com/1007-9327/full/v24/i26/2818.htm

Mihaljevic AL, Friess H, Schuhmacher C (2013) Clinical trials in gastric cancer and the future. J Surg Oncol 107(3):289–297

Pinto M, Oliveira C, Machado JC, Cirnes L, Tavares J, Carneiro F, Hamelin R, Hofstra R, Seruca R, Sobrinho-Simões M (2000) MSI-L gastric carcinomas share the hMLH1 methylation status of MSI-H carcinomas but not their clinicopathological profile. Lab Invest 80(12):1915–1923

Sharpe AH, Wherry EJ, Ahmed R, Freeman GJ (2007) The function of programmed cell death 1 and its ligands in regulating autoimmunity and infection. Nat Immunol 8:239–245. https://doi.org/10.1038/ni1443. [PMID: 17304234]

Sierzega M, Kaczor M, Kolodziejczyk P, Kulig J, Sanak M, Richter P (2017) Evaluation of serum microRNA biomarkers for gastric cancer based on blood and tissue pools profiling: the importance of miR-21 and miR-331. Br J Cancer 117:266–273. https://doi.org/10.1038/bjc.2017.190. [PMID: 28641313]

Siravegna G, Marsoni S, Siena S, Bardelli A (2017) Integrating liquid biopsies into the management of cancer. Nat Rev Clin Oncol 14:531–548. https://doi.org/10.1038/nrclinonc.2017.14. [PMID: 28252003]

Smyth EC, Wotherspoon A, Peckitt C, Gonzalez D, Hulkki-Wilson S, Eltahir Z, Fassan M, Rugge M, Valeri N, Okines A, Hewish M, Allum W, Stenning S, Nankivell M, Langley R, Cunningham D (2017) Mismatch repair deficiency, microsatellite instability, and survival: an exploratory analysis of the Medical Research Council Adjuvant Gastric Infusional

Chemotherapy (MAGIC) trial. JAMA Oncol 3:1197–1203. https://doi.org/10.1001/jamaoncol.2016.6762. [PMID: 28241187]

Suenaga M, Yamada S, Fuchs BC, Fujii T, Kanda M, Tanaka C, Kobayashi D, Fujiwara M, Tanabe KK, Kodera Y (2015) CD44 single nucleotide polymorphism and isoform switching may predict gastric cancer recurrence. J Surg Oncol 112:622–628. https://doi.org/10.1002/jso.24056. [PMID: 26416034]

Wagner AD, Unverzagt S, Grothe W, Kleber G, Grothey A, Haerting J, Fleig WE (2010) Chemotherapy for advanced gastric cancer. Cochrane Database Syst Rev (3):CD004064. https://doi.org/10.1002/14651858.CD004064.pub3. [PMID: 20238327]

Wu HH, Lin WC, Tsai KW (2014) Advances in molecular biomarkers for gastric cancer: miRNAs as emerging novel cancer markers. Expert Rev Mol Med 16:e1. https://doi.org/10.1017/erm.2013.16. [PMID: 24456939]

Zhang Y, Guan DH, Bi RX, Xie J, Yang CH, Jiang YH (2017) Prognostic value of microRNAs in gastric cancer: a meta-analysis. Oncotarget 8:55489–55510. https://doi.org/10.18632/oncotarget.18590. [PMID: 28903436]

Zhu X, Lv M, Wang H, Guan W (2014) Identification of circulating microRNAs as novel potential biomarkers for gastric cancer detection: a systematic review and meta-analysis. Dig Dis Sci 59:911–919. https://doi.org/10.1007/s10620-013-2970-9. [PMID: 24337687]

Chapter 5
Metabolic Markers for Early Detection of Gastrointestinal Cancers

A. M. V. N. Prathyusha, B. Prathap Naidu, and Pallaval Veera Bramhachari

Abstract Gastrointestinal cancers are a major threat to humans nowadays, due to their high incidence and mortality rates. The early detection of GI cancers was the key to prevention and treatment. Metabolomic profiling and analysis of volatile biomarkers are new and promising screening tools for the diagnosis of several cancers. Metabolites are small molecules produced during cell metabolism that represent the functional status of cell/tissue/malignant phenotype. Studies on biofluid metabolomics (serum, urine, plasma) are emerging fast on detecting new biomarkers for early diagnosis of cancer diseases and grasp a great promise for diagnostic applications. Volatile Organic Compounds (VOCs) signify diverse volatile metabolites which can be able to emit and diagnose in urine, breath, sweat, and feces. Nowadays, there is an increasing interest in the use and evaluation of VOCs with emerging analytical technologies in the diagnosis of GI cancer. With the advent of novel technologies in clinical diagnosis, metabolome analysis is an effective tool for metabolite profiling of biological processes in cells/tissues. This review emphasizes recent advancements in the identification of GI cancer biomarkers, particularly it focuses on metabolic markers and the emerging field of volatile biomarkers.

Keywords Gastrointestinal cancers · Metabolomic profiling · Volatile biomarkers · Clinical diagnosis

A. M. V. N. Prathyusha, B. Prathap Naidu, and Pallaval Veera Bramhachari equally contributed.

A. M. V. N. Prathyusha · B. P. Naidu · P. Veera Bramhachari (✉)
Department of Biotechnology, Krishna University, Machilipatnam, Andhra Pradesh, India

P. Veera Bramhachari, N. R. R. Neelapu (eds.), *Recent Advancements in Biomarkers and Early Detection of Gastrointestinal Cancers*, Diagnostics and Therapeutic Advances in GI Malignancies, https://doi.org/10.1007/978-981-15-4431-6_5

5.1 Introduction

Gastrointestinal (GI) cancers are malignant with high mortality rates in the GI tract and accessory organs of digestion (Tillu and Nagaraju 2017). GI malignancies account for 30% of all cancer incidence and represent approximately 40% of tumor-related mortality globally. However, GI cancers are declining worldwide; however, it is even higher in Eastern Asia, mainly in China and Japan (Prathyusha et al. 2017). GI cancers have a high rate of mortality and morbidity due to a lack of defined risk factors, delayed diagnoses in their early stages leading to the progression of advanced stages with high recurrence (Tillu and Nagaraju 2017). Diagnosis in earlier stages considerably increases the prognosis by 95% in GI cancer patients (Xu et al. 2013). Esophagus, pancreatic, colorectal cancer (CRC), and gastric adenocarcinoma are among the top GI cancers in both incidence and mortality.

Several diagnostic tests are available for screening cancer, viz. upper and lower GI endoscopy-biopsy with pathological evaluation of tissue (Correa 2013; Thrumurthy et al. 2013), non-endoscopy-based balloon cytology, barium esophagram (Groome et al. 2008), image-based tests such as CT (computerized tomography) scan, Cytosponge, MRI (magnetic resonance imaging), cystoscopy, sigmoidoscopy, colonoscopy (Wild et al. 2010) and serological markers (Zhang et al. 2013, Patel and Ahmed (2015). Carcinoembryonic antigen (CEA) and fecal occult blood testing (FOBT) are currently in use for diagnosis of CRC with relatively low sensitivities and specificities (Ransohoff and Sox 2016). Nonetheless, all pragmatic approaches possess specific limitations in diagnosing imaging techniques. A major drawback is in some cases cancers endure undetected until malignant stages, in turn, affect the survival rate. The accuracy of endoscopy-biopsy screening and examination depends on collecting appropriate tissue biopsy which is a common clinical practice while a battery of serological marker tests was unsuccessful for screening or surveillance because of their low accuracy, sensitivity, and specificity. Many biopsy techniques are deliberate to be uncomfortable, inconvenient to patients, and are often underutilized by large segments of the population for a variety of reasons, such as access and patient burden (Oakley-Girvan and Davis 2017). Nonetheless, this necessitates the requirement for cost-effective, sensitive, reliable, accurate, noninvasive, patient-friendly screening tools for GI cancer detection which are currently not employed. Ascertaining specific biomarkers for detection of GI cancer in the benign stage (at onset) to ensure patient survival seems very promising.

Metabolites are small molecules produced during cell metabolism that represent the functional status of cell/tissue/malignant phenotype. The uncontrolled proliferation and altered metabolisms (glycolysis, TCA cycle, choline, and fatty acid metabolism) are important hallmarks of the malignant phenotype. These altered metabolisms result in a metabolic landscape that diverges significantly among cancer and normal cells. The goal of metabolomics studies captures this unique metabolic portrait as a label of cancer. Furthermore, a wide range of metabolic by-products is produced, comprising of a battery of organic compounds characterized by volatility at body temperatures, thus known as volatile organic metabolites/compounds

(VOCs). Many VOCs are constantly produced in the body during various molecular pathways. VOCs are low molecular weight and volatile compounds detectable at low concentrations in biofluids, feces, and exhaled breath. Therefore, metabolites and VOCs can be highly informative regarding cell/tissue phenotype (Broza et al. 2018). Studies on biofluid metabolomics (serum, urine, plasma) and VOCs are emerging fast on detecting new biomarkers for early diagnosis of cancer diseases and grasp a great promise for diagnostic applications (Buck et al. 2017). Undoubtedly, several studies reported that specific regions of NMR and MS spectra differ among cancer patients and healthy controls (Patel and Ahmed 2015).

5.2 Metabolomic Analyses

Metabolic profiling involves two principal analytical procedures, viz. nuclear magnetic resonance (NMR) (Wang et al. 2017a, b) and mass spectrometry (MS) (Dettmer et al. 2007). Variations of NMR include high-resolution magic angle spinning (HR-MAS NMR) intended for tissue samples, ^{1}H-NMR, 2D-NMR (HSQC and HMQC) techniques have been widely used in NMR-based metabolomics marker study (Emwas 2015). The chemical and physical diversity of molecules in biofluids, and alterations in ionic strength, pH, temperature, etc. may impede NMR analysis (Zhang et al. 2013). MS may be coupled directly to a chromatographic column such as gas chromatography-mass spectroscopy (GC-MS), liquid chromatography-mass spectroscopy (LC-MS), or capillary electrophoresis (CE-MS) (García et al. 2017) based on appliance. However, LC-MS is further ramified into ultra-performance mass spectroscopy (UPLC-MS) (Chen et al. 2019) or high-performance mass spectroscopy (HPLC-MS) (Zhao et al. 2006), and LC/MS/MS (liquid chromatography with tandem mass spectrometry). MS analysis can be performed using different ionization methods and mass analyzers (quadrupole time-off light (Q-TOF), Fourier transform ion cyclotron resonance (FTICR), and Orbitrap) for separation of ions of different masses and detection of molecular fragments based on mass-to-charge ratios (m/z). HPLC, coupled with electrospray ionization and Q-TOF (HPLC/ESI/Q-TOF-MS), was widely in use and documented to be beneficial clinically for early diagnosis of GI cancer due to precise qualitative analysis (Callejón-Leblic et al. 2016; Wang et al. 2017a, b).

Metabolome analysis was categorized into three major ways: targeted analysis, metabolic fingerprinting, and metabolite profiling (untargeted analysis). Untargeted analysis (metabolic profiling) usually emphasizes the measurement of levels of metabolites of all detectable metabolites in a given sample. NMR technique is widely in use for metabolomics fingerprinting (Emwas 2015). Metabolic fingerprinting reflects total metabolite fingerprint as a rare pattern illustrating a glimpse of the particular cell line or tissue metabolism. The targeted analysis emphasizes on the quantification and documentation of metabolites, of a specific metabolic pathway or direct product of drug administering or food intake. The metabolites involved in the study are usually predetermined, in targeted analysis, and the sample preparation is modified to diminish interference from other metabolites (Zhang et al. 2013). Djukovic and group employed targeted metabolite procedure to examine the role of nucleosides as cancer biomarkers in EAC (Esophageal Adenocarcinoma) using a serum-NMR technique (Djukovic et al. 2010).

Two main approaches have been developed that are used to assess VOCs in biological samples. Identification of the VOC spectrum allows one to study each compound separately and estimate the likelihood that it can serve as an individual biomarker of disease. One approach is based on analytical techniques such as GC-MS (Altomare et al. 2013), proton transfer reaction mass spectrometry (PTR-MS) (Jordan et al. 1995) and selected ion flow tube mass spectrometry (SIFT-MS) (Krilaviciute et al. 2015). Furthermore, the second approach is based on diverse chemically sensitive gas sensors that provide a total assessment of the mixture/profile of VOCs in a sample (Lourenço and Turner 2014).

Complex metabolite profiles obtained after analytical techniques were interpreted and characterized by multivariate pattern recognition statistical analysis, viz. (1) principle component analysis (PCA) (Wang et al. 2018), consensus PCA (CPCA), partial least square discriminant analysis (PLS-DA) (Nakajima et al. 2018), orthogonal signal correction (OSC), orthogonal partial least square discriminant analysis (OPLS-DA) (Li et al. 2013), and graded clustering analysis (GCA) (Wishart et al. 2016; Lai et al. 2018). XCMS is the most commonly used tool for metabolite profiling to analyze MS data by matching and nonlinear peak alignment. SIMCA-P is employed for multivariate data analysis (MVDA) by PLS, OPLS, and PCA analysis. MZmine 2 is a modular framework for visualizing, analyzing, and processing MS data. MetaboAnalyst is a web-based pipeline for statistical analysis, metabolomic data processing, and consequent functional elucidation. Notably, several additional tools also help in spectrometric data analysis. Davis et al. employed Mann–Whitney statistical analysis tool to correlate individual metabolite concentration among controls and EC patients (Davis et al. 2012).

5.3 Serum Metabolomics Signatures

Digestive tract malignant neoplasms are the most common reason for cancer-related mortality globally. Ikeda and hid co-workers reported more sensitive metabolites for diagnosis of GI cancer using analytical tools LC-MS, GC-MS, and NMR followed by PCA and PLS-DA statistical analysis. Their studies reported the metabolic differences among esophagus, gastric, and CRC cancer signifying the importance of different metabolisms in these cancers. Perhaps, these studies using multiple classifications analysis identify metabolite alteration levels namely malonic acid and L-serine is characteristic for EC. Alterations in concentrations of pyruvic acid and 3-hydroxy propionic acid is characteristic for gastric cancer while L-glutamine and glucuronic lactone signify colorectal cancer (Ikeda et al. 2012).

5.3.1 Esophagus and Stomach

Esophagus cancers are extremely malignant affecting the upper digestive tract with a 5-year survival rate in less than 15%. Zhang and colleagues (2013) analyzed the serum metabolic signatures of EC patients with healthy control using ^{1}H NMR and UPLC based focused metabolic profiling. EC cells exhibited significant perturbations in glucose (reduced glucose and increased lactate levels in EC patients), lipids (apolipoproteins, LDL, VLDL, and unsaturated lipid were considerably reduced in the serum of EC patients), energy (serum creatinine and creatine concentrations were considerably elevated among EC patients), and amino acid metabolism. However, 12 biomarkers using NMR and 7 markers from UHPLC focused metabolomics were identified. Strikingly, the ketone bodies, viz. acetoacetate and β-hydroxybutyrate level were specifically elevated in EC patients' serum suggesting a promoted β-oxidation. Amino acids (L-Tryptophan, L-Tyrosine, linoleic acid, and palmitic acid) have been identified to be modified in ESCC (Zhang et al. 2013). Wang et al. reported serum metabolites for the early-stage detection and discrimination among different stages of esophageal squamous cell carcinoma (ESCC) using nonparametric Kruskal–Wallis rank-sum test. Metabolites of glycerophospholipid, linoleic acid, and choline metabolism were to be found dysregulated in the ESCC population assessed to healthy individuals. Decreased trends of three biomarkers, namely dodecanoic acid, lysophosphatidic acid (LPA), and 4-lysophosphatidylcholines (LysoPC), reported being clear hallmarks for ESCC progression (Wang et al. 2016).

5.3.2 Colorectal Cancer (CRC)

CRC is one of the most prevailing cancer types globally with high morbidity and mortality. Uchiyama et al. reported serum metabolites for early-stage detection. Metabolome analysis was performed by CE-TOFMS for serum samples of CRC patients and controls followed by HCA and PCA statistical analysis and detected metabolites were plotted using VANTED software. Seventeen metabolites were reported to be correlated with CRC by elevation and 16 by downregulation in patients with adenoma in comparison with controls. Clustering identified upregulation of 7 and downregulation of 29 metabolites in stage I CRC, while upregulation of 11 and downregulation of 19 metabolites in stage II CRC. In stage III CRC, 5 metabolites reported to be upregulated whereas 39 downregulated; and in stage IV, 6 metabolites reported to be upregulated and 28 downregulated compared to normal controls. 3-hydroxybutyric acid was elevated in CRC stages and isovaleric acid, ornithine, benzoic acid, and the amino acids His, Lys, and Trp were downregulated. Of these, decanoic acid, octanoic acid (upregulated), histidine (downregulated), and benzoic acid exhibit significant correlation with all CRC stages. Benzoic acid is one of the metabolites that originated from procyanidins degradation by human gut microbiota. Benzoic acid was reported to be an excellent diagnostic marker to detect early and all the stages of CRC (Uchiyama et al.2017). Hata and his co-workers reported a promising serum biomarker GTA-446 (gastrointestinal tract acids) for primary CRC screening with the aim of early detection and to detect peoples at higher risk of CRC (Hata et al. 2017).

5.4 Plasma Metabolic Signatures

Plasma free amino acids (PFAA) are one of the most essential compounds for focused metabolomics as they play important physiological roles as elementary metabolites and metabolic regulators. Several studies reported that amino-acid levels were drastically declined in early-stage cancer populations, irrespective of subsequent progression. Particularly, significant reductions in the concentration of all amino acids were detected in both GC and CRC patients. Glutamine, tryptophan, histidine, proline, and ornithine amino acids were related to all types of cancer unveiled by univariate analysis (Miyagi et al. 2011).

5.4.1 Colorectal Cancer (CRC)

Cancers are characterized by diverse metabolic phenotypes due to alterations in key metabolic pathways, viz. glycolysis or tricarboxylic acid (TCA) cycle, etc. Geijsen et al. (2019) reported the correlation between plasma metabolites and colorectal cancer. Wang et al. reported a plasma metabolite marker for CRC diagnosis using a two-stage case-control study employing UPLC/Q-TOF MS/MS and PCA analysis. Glycochenodeoxycholate, L-Tryptophan, 13-OxoODE, IDP, and LysoPC (16:0), a five-biomarker panel, exhibited excellent diagnostic performance, favorable biological significance, and tumor specificity in Northeast China population. A six-biomarker panel of L-Phenylalanine, Linoleic acid, Citric acid, Inosine, Glycocholic acid, and LysoPC (14:0) showed the best extrapolation in diverse populations. Four metabolites (L-Tryptophan, Linoleic acid, Glycocholic acid, and LysoPC (16:0)) were ultimately recommended as the best combination with multiple advantages (Wang et al. 2018).

5.4.2 Esophagus and Stomach

In recent years, metabolomics studies on EC were performed by diverse analytical techniques to discover potential therapeutic markers. Liu et al. reported the metabolic profiling of ESCC by using UPLC-ESI-TOFMS and identified six upregulated molecules, namely phosphatidic acid, phosphatidylserine, phosphatidylethanolamine phosphatidylinositol, sphinganine 1-phosphate, and phosphatidylcholine.

Furthermore, bile acid, lithocholic acid taurine conjugate reported being significantly upregulated whereas 5-β-cyprinol sulfate and desmosine/isodesmosine are downregulated in the plasma of ESCC patients (Liu et al. 2013).

5.5 Urinary and Fecal Metabolic Signatures

Urine is a biofluid usually used by metabolomics scientists, due to the collection of large volumes and patient-friendly.

5.5.1 Colorectal Cancer

Urine metabolic profiles were analyzed by Wang and his co-workers using ^{1}H NMR along with OPLS-DA statistical validation by permutation analysis in comparison with healthy controls at stage I/II. They reported perturbation in amino acid (asparagine, alanine, cysteine, and phenylalanine), glycolysis, TCA cycle, choline (acetoacetate, guanidinoacetate), vitamin-B3 (Trigonelline), and urea metabolisms. Metabolites upregulated and downregulated at different stages were summarized in Table 5.1 (Wang et al. 2017a, b). Le Gall and his research group quantified fecal extracts from CRC patients and screened over 80 molecules using NMR. Furthermore, isobutyrate, branched-chain fatty acids (BCFA), valerate, isovalerate, and phenylacetate were reported to be upregulated while reduced concentrations of amino acids, methanol, sugars, and bile acids (lithodeoxycholate, cholate, and deoxycholate) in the fecal extracts of ESCC patients (Le Gall et al. 2019).

5.5.2 Esophagus and Stomach

Altered carbohydrates, amino acids, lipid, and ketone metabolisms were reported to be hallmarks for EC. Xu and his research group reported global urine metabolic profile of ESCC patients and healthy controls using LC-MS along with MVDA and reported 19 potential biomarkers related to perturbations of amino acids fatty acid β-oxidation and nucleotide metabolism. Metabolites upregulated and downregulated at different stages were summarized in Table 5.1 (Xu et al. 2016). Recently, studies are employing metabolomics technology to tissue, plasma, serum, and urine samples revealed alterations in choline, glucose, amino acid, fatty acid, linoleic acid, and energy metabolism in EC cells. These metabolites upregulation and downregulation reflected in blood samples of patients indicates the importance of these metabolisms in esophagus cancer progression and early detection.

Table 5.1 An overview of study design, analytical techniques, identified compounds for early detection of GI cancers

S. no	Cancer type	Analytical platform	Sample type	Statistical analysis	Metabolites of cancer compared to control		Biomarker	References
					Upregulated	Downregulated		
1.	CRC	CE-TOFMS	Serum		Decanoic acid Octanoic acid	Histidine, Benzoic acid (biomarker for early CRC)	Benzoic acid	Uchiyama et al. (2017)
2.	CRC	^{1}H NMR	Fecal water extracts	Robust PCA	Cysteine, proline	Acetate and butyrate		Monleon et al. (2009)
3.	CRC	UPLC/Q-TOF-MS/MS	Plasma	PCA	Phosphatidylcholines, linoleic acid, gamma-linolenic acid, 13-L-hydroperoxylinoleic acid, 13S-hydroxyoctadecadienoic acid, and 13-OxoODE	2-Aminobenzoic acid, 13-OxoODE, Citric acid, 2′-Deoxyinosine triphosphate, Taurocholic acid, and LysoPC	L-Tryptophan, Linoleic acid, Glycocholic acid, and LysoPC	Wang et al. (2018)
4.	CRC	^{1}H NMR	Urine	OPLS-DA	Thymidine, Fumarate, Hippurate, Cis-Aconitate, Pyridoxic acid, Cinnamic acid, Homogentisic acid, Indoleacetate, Trigonelline	Creatine, Creatinine, Uracil, Urea		Wang et al. (2017a, b)
5.	Esophageal cancer	NMR-based metabolomics and UHPLC-based focused metabolomics	Blood serum	OPLS-DA, PCA	β-Hydroxybutyrate acetoacetate Creatine Creatinine, Lactate, Glutamate, Glutamine, Histidine	LDL/VLDL, Unsaturated lipids, Acetate, Glucose, Tyrosine		Zhang et al. (2013)
6.	ESCC	UHPLC-QTOF/MS	Blood serum	PCA, PLS-DA		Dodecanoic acid, lysophosphatidic acid (LPA), and 4-lysophosphatidylcholines (LysoPC)		Wang et al. (2016)

7.	ESCC	^{1}H NMR	Serum	PCA, PLS-DA			α-Glucose, choline, glutamine, glutamate, valine, and dihydrothymine	Yang et al. (2019)
8.	Esophageal cancer	^{1}H NMR	Tissue	OPLS-DA, PCA	Acetate, short-chain fatty acid and GABA, creatinine, creatine, DMG, DMA, and TMA	Glucose, AMP and NAD, upregulation of formate		Wang et al. (2013)
9.	ESCC	UPLC/TOF/MS	Plasma	PCA	Phospholipids, sphinganine 1-phosphate, bile acid, lithocholic acid taurine conjugate	Desmosine/isodesmosine and 5-β-cyprinol sulfate		Liu et al. (2013)
10.	GI cancer	LC/MS, NMR, GC/MS	Serum	PCA, PLS-DA			Malonic acid and L-serine -esophageal cancer 3-hydroxypropionic acid and pyruvic acid- gastric cancer L-alanine, glucuronoic lactone and L-glutamine-CRC	Ikeda et al. (2012)
11.	ESCC	LC-MS/MS, NMR	Urine	OPLS-DA	Pyroglutamic acid, Indoxyl, Urocanic acid, L-Carnitine, L-Fucose, Uric acid, Acetylcarnitine, Deoxycytidine, cAMP, cGMP Phenylacetylglutamine	Paraxanthine, Heptanoylcarnitine, Octenoylcarnitine,. Nonenoylcarnitine, Nonanoylcarnitine, Decanoylcarnitine, Undecenoylcarnitine, Undecanoylcarnitine		Xu et al. (2016)

(continued)

Table 5.1 (continued)

S. no	Cancer type	Analytical platform	Sample type	Statistical analysis	Metabolites of cancer compared to control		Biomarker	References
					Upregulated	Downregulated		
12.	CRC	MRB-CE-ESI-MS	Urine		Isoleucine, valine, arginine, lactate acid, and leucine	Histidine, methionine, serine, aspartic acid, citric acid, succinate, and malic acid		Chen et al. (2012)
13.	CRC	LS–MS/MS	Serum			GTA-446	GTA-446	Hata et al. (2017)
14.	CRC	RRLC-TOF/MS	Urine	PCA, PLS-DA		Dihydrosphingosine, Sphinganine, O-octanoyl-R-carnitine		Yue et al. (2013)
15.	CRC	^{1}H NMR	Fecal extracts		Branched-chain fatty acids (BCFA), isovalerate and isobutyrate, valerate, and phenylacetate	Amino acids, sugars, methanol and bile acids (deoxycholate, lithodeoxycholate, and cholate		Le Gall et al. (2019)
16.	CRC	LC-MS	Urine	PLS-DA			Polyamines	Nakajima et al. (2018)
17.	CRC	DI-ESI-FTICR MS a	Serum	OPLS-DA			Palmitic amide, oleamide, hexadecanedioic acid, octadecanoic acid, eicosatrienoic acid, LPC (18:2), LPC (20:4), LPC (22:6), myristic acid, and LPC (16:0)	Li et al. (2013)

5.6 Volatile Organic Compounds

VOCs are a diverse group of carbon compounds emitted from the human body that reflects the metabolic condition of the person detected in blood, breath, and excreted body fluids. Hundreds of different VOCs have been reported in cancer patients from different sources. Acids, ketones, alcohols, amines, aldehydes, O-heterocycles, N-heterocycles, VOCs, and other VSCs have been detected and quantified. However, concentrations of specific organic compounds in the exhaled breath of cancer patients are thought to be the result of oxidative stress, abnormal metabolic processes, or inability of biological systems to detoxify ROS (reactive oxygen species). Some VOCs are thought to originate from cancer cells as a result of abnormal metabolism that does not exist in healthy people (Oakley-Girvan and Davis 2017). Among several extraction procedures SPME (solid-phase microextraction) is a prevailing technique for sample preparation, sampling a wide range of analytes in breath and biological samples (Kim et al. 2019).

The four studies targeting to identify VOC in breath samples of CRC patients suggested diverse patterns of potential biomarkers were not completely matched, with an exemption of 1,3-dimethyl benzene reported by both Peng and Altomare groups individually (Peng et al. 2010; Altomare et al. 2013). Furthermore, 4-methyl octane was reported as a potential biomarker both by Altomare and Amal groups individually (Altomare et al. 2013; Amal et al. 2016). These diverse patterns possibly due to the involvement of different methods in sample collection and analysis. Wang et al. examined two different samples, of CRC patients i.e., exhaled breath and blood using SPME-GC-MS technique. Two different VOC patterns that did not fit each other probably due to different sample characteristics are reported. Remarkably, these authors identified reduced levels of 6-t-butyl-2,2,9,9-tetramethyl-3,5-decadien-7-yne in breath and blood sample. Kumar and his research group analyzed breath samples from 81 esophageal or gastric adenocarcinoma patients against 129 controls including benign upper gastrointestinal diseases, Barrett's metaplasia, and normal upper GI tract. Twelve VOCs—hexanoic acid, pentanoic acid, methyl phenol, ethyl phenol, phenol, butanal, hexanal, pentanal, octanal, heptanal, decanal, and nonanal, were reported at extremely high levels in cancer groups compared to controls (Kumar et al. 2015). Kim and his co-workers employed SPME and 2D GC-MS to analyze 30 random plasma samples from CRC patients and reported five VOCs, among them 2,3,4-trimethylhexane (reduced) and 2,4-dimethylhept-1-ene (enhanced) were both lipid peroxidation products (Kim et al. 2019). Wang and his co-workers analyzed blood VOCs of SW620 CRC mice for a course of 12 and 26 days. They reported eight VOCs which progressively augmented with tumor growth (acetone, glycerol, arsenous acid, 2-heptanone, 4-nonanone, tris (trimethylsilyl) ester, butylated hydroxytoluene, l-alanine ethylamide (S)-, and 3-heptanone,4-methyl 2-dodecanone). The concentration of VOCs peaked at maximum tumor size and decreased promptly after tumor resection. Cyclotrisiloxane, hexamethyl was utilized by the tumor, and its concentration progressively reduced with the size of the tumor (Wang et al. 2019).

5.7 Nanomaterial-Based Breath Tests

Nanomaterial-based breath tests were first developed by Haick and co-workers that could categorize among different cancer types in exhaled breath, regardless of patients' lifestyle, habits, gender, and other confounding factors. Xu et al. used GC-MS for chemical analysis of breath samples reported five VOCs (2-butoxy-ethanol, 2-propenenitrile, furfural, isoprene, and 6-methyl-5-hepten-2-one) elevated in gastric cancer and gastric ulcer patients. Three different DFA models were developed to achieve excellent discrimination among them (1) GC vs. benign gastric conditions, reported with 89% sensitivity; 90% specificity; (2) different stages of GC early-stage (I and II) and late-stage (III and IV), with 89% sensitivity; 94% specificity (Xu et al. 2013). Tong et al. reported four nanomaterial-based biomarkers in gastric cancer patients vs. healthy individuals 1,3-Dioxolan-2-one, Undecane, 2,3-Butanediol, 3,8-dimethyl- *N,N*-Dimethylacetamide Hexadecane while (*p*-hydroxyphenyl), 1,3-Dioxolane-2-methanol and 3,5-Decadien-7-yne, Phosphonic acid, 6-t-butyl-2,2,9,9-tetramethyl-1,6-Dioxacyclododecane-7,12-dione for carcinoma patients vs. gastric ulcer patients (Tong et al. 2017).

5.8 Integration of Metabolomic Networks

In the human body, most cancer cells exhibit hypoxic environment due to increased cell division at early stages resulting in increased aerobic and anaerobic glycolysis which is accompanied by disrupted TCA cycle and even other metabolisms creating a mutagenic phenotype. Tumor cells predominantly generate energy through glycolysis rather than oxidative phosphorylation via the TCA cycle, even in aerobic conditions (Warburg effect) (Armitage and Southam 2016). This results in the production of lactate in large amounts by the tumor and the anabolic synthesis of nucleotides, amino acids, and lipids. Lactate produced in large amounts is converted to glucose in liver via Cori cycle ends in acetyl Co-A accumulation and citrate upregulation (disruption of TCA cycle). If acetyl-CoA is not well utilized in TCA cycle, ketogenesis takes place and thus the accumulation of ketone bodies, viz. acetoacetate and β-hydroxybutyrate (Zhang et al. 2013). Increased glucose utilization favors energy metabolism, NADPH recycling to maintain serum creatine, creatinine, and glutathione levels for an optimal cellular redox status, respectively (Wang et al. 2013). Many blood amino acids downregulated in cancer patients indicate an increased demand for and overutilization of amino acids in tumor tissue. Altered fatty acid metabolism is also reported in cancer patient's plasma, by reduced levels of several unsaturated lipids, VLDL, and LDL (Hasim et al. 2012). Wang et al. first reported the importance of essential fatty acids (linolenic acid) in proliferating tumor cells for the biosynthesis of prostaglandins and cell-membrane assembly and their metabolites as biomarkers (Wang et al. 2018). In summary, altered/disrupted metabolomic pathways comprise fluctuations in glycolysis, Cori cycle, TCA cycle, an amino acid (Alanine, Leucine, Valine, Isoleucine, 1-methylhistidine, and Glycoprotein), a ketone body, fatty acid (unsaturated lipids, VLDL, and LDL), energy metabolisms, and urea cycle. Metabolites accumulate in these pathways represent typical metabolic and VOC signatures in GI cancer patients.

5.9 Conclusion and Future Directions

Currently available techniques for diagnosis and surveillance of GI cancers are expensive, invasive, and not suitable for early detection. Robust, low-cost, and noninvasive biomarkers to facilitate early screening, surveillance are scanty, indicating the necessity for the development of more efficient methods. In this sense, the metabolomics field emerged highly promising in the past few years, with large advancements in GI cancer diagnosis based on metabolites. Metabolic profiling has been used to find novel biomarkers for early detection of cancers. However, metabolomic analysis is a promising approach in cancer diagnosis, with certain constraints comprising necessity to evaluate existing metabolites, data redundancy, false discovery problems, and cost limitations endure major hurdles for metabolomics research. Issues related to sensitivity, specificity, and accuracy for biomarker detection must be dealt with to unleash metabolomics potential. Furthermore, the diagnostic accuracy of metabolic signatures needs to be established (Figs. 5.1 and 5.2).

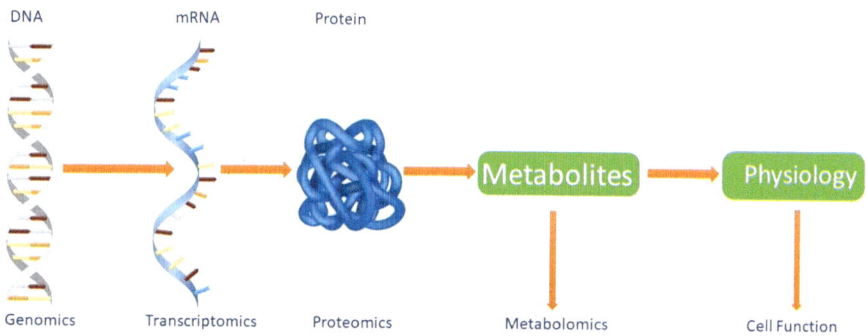

Fig. 5.1 Genomics to metabolomics

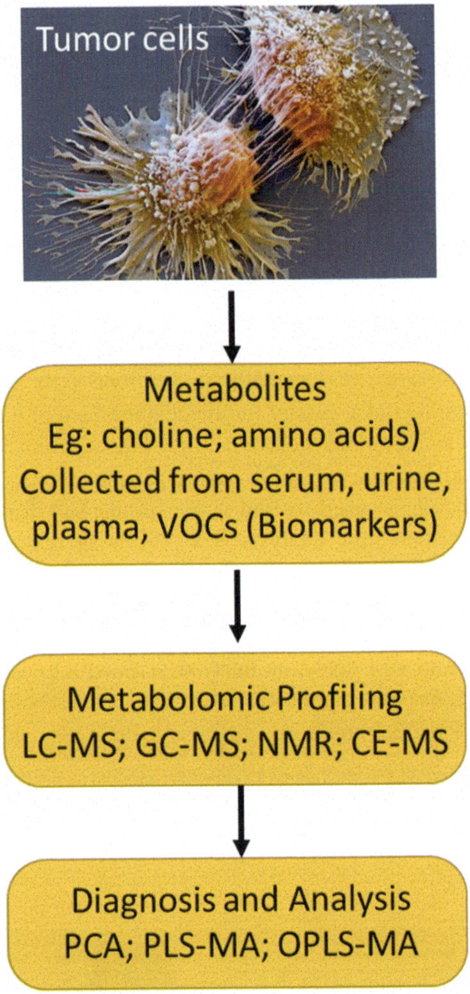

Fig. 5.2 Schematic representation of metabolome profiling

Acknowledgments The authors gratefully acknowledge UGC, Government of India, for financial support to AMVNP in the form of UGC-CSIR-JRF & SRF under the Grant No: F.No. 19-1/2015 (SA-1).

Conflict of Interest We declare we do not have any conflict of interest.

References

Altomare DF, Di Lena M, Porcelli F, Trizio L, Travaglio E, Tutino M et al (2013) Exhaled volatile organic compounds identify patients with colorectal cancer. Br J Surg 100(1):144–150

Amal H, Leja M, Funka K, Lasina I, Skapars R, Sivins A et al (2016) Breath testing as potential colorectal cancer screening tool. Int J Cancer 138(1):229–236

Armitage EG, Southam AD (2016) Monitoring cancer prognosis, diagnosis and treatment efficacy using metabolomics and lipidomics. Metabolomics 12(9):146

Broza YY, Vishinkin R, Barash O, Nakhleh MK, Haick H (2018) Synergy between nanomaterials and volatile organic compounds for non-invasive medical evaluation. Chem Soc Rev 47 (13):4781–4859

Buck A, Aichler M, Huber K, Walch A (2017) In situ metabolomics in cancer by mass spectrometry imaging. In: Advances in cancer research, vol 134. Academic, Cambridge, pp 117–132

Callejón-Leblic B, García-Barrera T, Grávalos-Guzmán J, Pereira-Vega A, Gómez-Ariza JL (2016) Metabolic profiling of potential lung cancer biomarkers using bronchoalveolar lavage fluid and the integrated direct infusion/gas chromatography mass spectrometry platform. J Proteome 145:197–206

Chen JL, Fan J, Yan LS, Guo HQ, Xiong JJ, Ren Y, Hu JD (2012) Urine metabolite profiling of human colorectal cancer by capillary electrophoresis mass spectrometry based on MRB. Gastroenterol Res Prac 2012:8

Chen J, Hou H, Chen H, Luo Y, Zhang L, Zhang Y et al (2019) Urinary metabolomics for discovering metabolic biomarkers of laryngeal cancer using UPLC-QTOF/MS. J Pharm Biomed Anal 167:83–89

Correa P (2013) Gastric cancer: overview. Gastroenterol Clin N Am 42:211–217

Davis VW, Schiller DE, Eurich D, Sawyer MB (2012) Urinary metabolomic signature of esophageal cancer and Barrett's esophagus. World J Surg Oncol 10(1):271

Dettmer K, Aronov PA, Hammock BD (2007) Mass spectrometry-based metabolomics. Mass Spectrom Rev 26(1):51–78

Djukovic D, Baniasadi HR, Kc R, Hammoud Z, Raftery D (2010) Targeted serum metabolite profiling of nucleosides in esophageal adenocarcinoma. Rapid Commun Mass Spectrom 24 (20):3057–3062

Emwas AHM (2015) The strengths and weaknesses of NMR spectroscopy and mass spectrometry with particular focus on metabolomics research. In: Metabonomics. Humana Press, New York, pp 161–193

García A, Godzien J, López-Gonzálvez Á, Barbas C (2017) Capillary electrophoresis mass spectrometry as a tool for untargeted metabolomics. Bioanalysis 9(1):99–130

Geijsen AJ, Brezina S, Keski-Rahkonen P, Baierl A, Bachleitner-Hofmann T, Bergmann MM et al (2019) Plasma metabolites associated with colorectal cancer: a discovery-replication strategy. Int J Cancer 145(5):1221–1231

Groome M, Lindsay J, Ross PE, Cotton JP, Hupp TR, Dillon JF (2008) Use of oesophageal stress response proteins as potential biomarkers in the screening for Barrett's oesophagus. Eur J Gastroenterol Hepatol 20(10):961–965

Hasim A, Ma H, Mamtimin B, Abudula A, Niyaz M, Zhang LW et al (2012) Revealing the metabonomic variation of EC using 1 H-NMR spectroscopy and its association with the clinicopathological characteristics. Mol Biol Rep 39(9):8955–8964

Hata T, Takemasa I, Takahashi H, Haraguchi N, Nishimura J, Hata T et al (2017) Downregulation of serum metabolite GTA-446 as a novel potential marker for early detection of colorectal cancer. Br J Cancer 117(2):227

Ikeda A, Nishiumi S, Shinohara M, Yoshie T, Hatano N, Okuno T et al (2012) Serum metabolomics as a novel diagnostic approach for gastrointestinal cancer. Biomed Chromatogr 26(5):548–558

Jordan A, Hansel A, Holzinger R, Lindinger W (1995) Acetonitrile and benzene in the breath of smokers and non-smokers investigated by proton transfer reaction mass spectrometry (PTR-MS). Int J Mass Spectrom Ion Process 148(1–2):L1–L3

Kim S, Yin X, Prodhan MAI, Zhang X, Zhong Z, Kato I (2019) Global plasma profiling for colorectal cancer-associated volatile organic compounds: a proof-of-principle study. J Chromatogr Sci 57(5):385–396

Krilaviciute A, Heiss JA, Leja M, Kupcinskas J, Haick H, Brenner H (2015) Detection of cancer through exhaled breath: a systematic review. Oncotarget 6(36):38643

Kumar S, Huang J, Abbassi-Ghadi N, Mackenzie HA, Veselkov KA, Hoare JM et al (2015) Mass spectrometric analysis of exhaled breath for the identification of volatile organic compound biomarkers in esophageal and gastric adenocarcinoma. Ann Surg 262(6):981–990

Lai Z, Tsugawa H, Wohlgemuth G, Mehta S, Mueller M, Zheng Y et al (2018) Identifying metabolites by integrating metabolome databases with mass spectrometry cheminformatics. Nat Methods 15(1):53

Le Gall G, Guttula K, Kellingray L, Tett AJ, ten Hoopen R, Kemsley EK et al (2019) Correction: metabolite quantification of faecal extracts from colorectal cancer patients and healthy controls. Oncotarget 10(17):1660

Li F, Qin X, Chen H, Qiu L, Guo Y, Liu H et al (2013) Lipid profiling for early diagnosis and progression of colorectal cancer using direct-infusion electrospray ionization Fourier transform ion cyclotron resonance mass spectrometry. Rapid Commun Mass Spectrom 27(1):24–34

Liu R, Peng Y, Li X, Wang Y, Pan E, Guo W et al (2013) Identification of plasma metabolomic profiling for diagnosis of esophageal squamous-cell carcinoma using an UPLC/TOF/MS platform. Int J Mol Sci 14(5):8899–8911

Lourenço C, Turner C (2014) Breath analysis in disease diagnosis: methodological considerations and applications. Metabolites 4(2):465–498

Miyagi Y, Higashiyama M, Gochi A, Akaike M, Ishikawa T, Miura T et al (2011) Plasma free amino acid profiling of five types of cancer patients and its application for early detection. PLoS One 6(9):e24143

Monleon D, Morales JM, Barrasa A, Lopez JA, Vazquez C, Celda B (2009) Metabolite profiling of fecal water extracts from human colorectal cancer. NMR Biomed 22(3):342–348

Nakajima T, Katsumata K, Kuwabara H, Soya R, Enomoto M, Ishizaki T et al (2018) Urinary polyamine biomarker panels with machine-learning differentiated colorectal cancers, benign disease, and healthy controls. Int J Mol Sci 19(3):756

Oakley-Girvan I, Davis SW (2017) Breath based volatile organic compounds in the detection of breast, lung, and colorectal cancers: a systematic review. Cancer Biomark 21(1):29–39

Patel S, Ahmed S (2015) Emerging field of metabolomics: big promise for cancer biomarker identification and drug discovery. J Pharm Biomed Anal 107:63–74

Peng G, Hakim M, Broza YY, Billan S, Abdah-Bortnyak R, Kuten A et al (2010) Detection of lung, breast, colorectal, and prostate cancers from exhaled breath using a single array of nanosensors. Br J Cancer 103(4):542

Prathyusha AMVN, Raghu G, Bramhachari PV (2017) HIF-1α: its role in metastasis of oesophageal malignancy. In: Role of transcription factors in gastrointestinal malignancies. Springer, Singapore, pp 71–87

Ransohoff DF, Sox HC (2016) Clinical practice guidelines for colorectal cancer screening: new recommendations and new challenges. JAMA 315(23):2529–2531

Thrumurthy SG, Chaudry MA, Hochhauser D, Mughal M (2013) The diagnosis and management of gastric cancer. BMJ 347:f6367

Tillu H, Nagaraju GP (2017) YY1 and KLF4: their role in gastrointestinal malignancies. In: Role of transcription factors in gastrointestinal malignancies. Springer, Singapore, pp 3–15

Tong H, Wang Y, Li Y, Liu S, Chi C, Liu D et al (2017) Volatile organic metabolites identify patients with gastric carcinoma, gastric ulcer, or gastritis and control patients. Cancer Cell Int 17 (1):108

Uchiyama K, Yagi N, Mizushima K, Higashimura Y, Hirai Y, Okayama T et al (2017) Serum metabolomics analysis for early detection of colorectal cancer. J Gastroenterol 52(6):677–694

Wang L, Chen J, Chen L, Deng P, Xiang P, Li M et al (2013) 1 H-NMR based metabonomic profiling of human esophageal cancer tissue. Mol Cancer 12(1):1

Wang J, Zhang T, Shen X, Liu J, Zhao D, Sun Y et al (2016) Serum metabolomics for early diagnosis of esophageal squamous cell carcinoma by UHPLC-QTOF/MS. Metabolomics 12 (7):116

Wang Z, Lin Y, Liang J, Huang Y, Ma C, Liu X, Yang J (2017a) Nmr-based metabolomic techniques identify potential urinary biomarkers for early colorectal cancer detection. Oncotarget 8(62):105819

Wang D, Li W, Zou Q, Yin L, Du Y, Gu J, Suo J (2017b) Serum metabolomic profiling of human gastric cancer and its relationship with the prognosis. Oncotarget 8(66):110000

Wang M, Long Z, Tian J, Chen S, Sun H, Jian R, et al (2018) Plasma biomarkers for colorectal cancer diagnosis based on metabolomics

Wang G, Li Y, Liu M, Guo N, Han C, Liu D et al (2019) Determination of volatile organic compounds in SW620 colorectal cancer cells and tumor-bearing mice. J Pharm Biomed Anal 167:30–37

Wild N, Andres H, Rollinger W, Krause F, Dilba P, Tacke M, Karl J (2010) A combination of serum markers for the early detection of colorectal cancer. Clin Cancer Res 16(24):6111–6121

Wishart D, Mandal R, Stanislaus A, Ramirez-Gaona M (2016) Cancer metabolomics and the human metabolome database. Metabolites 6(1):10

Xu ZQ, Broza YY, Ionsecu R, Tisch U, Ding L, Liu H et al (2013) A nanomaterial-based breath test for distinguishing gastric cancer from benign gastric conditions. Br J Cancer 108(4):941

Xu J, Chen Y, Zhang R, He J, Song Y, Wang J et al (2016) Global metabolomics reveals potential urinary biomarkers of esophageal squamous cell carcinoma for diagnosis and staging. Sci Rep 6:35010

Yang Z, Liu Y, Ma L, Wen X, Ji H, Li K (2019) Exploring potential biomarkers of early stage esophageal squamous cell carcinoma in pre-and post-operative serum metabolomic fingerprint spectrum using 1H-NMR method. Am J Transl Res 11(2):819

Yue H, Wang Y, Zhang Y, Ren H, Wu J, Ma L, Liu S (2013) A metabonomics study of colorectal cancer by rrlc-qtof/ms. J Liq Chromatogr Relat Technol 36(4):428–438

Zhang X, Xu L, Shen J, Cao B, Cheng T, Zhao T et al (2013) Metabolic signatures of esophageal cancer: NMR-based metabolomics and UHPLC-based focused metabolomics of blood serum. Biochim Biophys Acta 1832(8):1207–1216

Zhao X, Wang W, Wang J, Yang J, Xu G (2006) Urinary profiling investigation of metabolites with cis-diol structure from cancer patients based on UPLC-MS and HPLC-MS as well as multivariate statistical analysis. J Sep Sci 29(16):2444–2451

Chapter 6
Current Status of MicroRNA-Based Biomarkers for Gastric Cancer

Prakash C. Sharma and Renu Verma

Abstract Gastric cancer (GC) is a heterogeneous disease and remains one of the leading causes of cancer-related mortalities worldwide. The management of the disease is difficult due to late diagnosis and poor response to available treatment regimes. Currently available gastric cancer biomarkers have serious limitations in their applicability in diagnosis and prognosis of the disease. Therefore, potential biomarkers, particularly with noninvasive assays, are urgently required for the early detection and efficient prediction of therapeutic response and prognosis of gastric cancer. MicroRNAs (miRNAs) are a class of small non-coding RNA sequences that play an important role in modulating key biological processes by regulating the expression of target genes. These molecules are abnormally expressed within the tumor tissues and associated biological fluids including blood, gastric juice, and urine of GC patients. Recent experimental findings have led to the identification of a large number of miRNAs implicated in the occurrence and progression of gastric cancer. miRNAs contribute to gastric carcinogenesis by regulating the expression of different oncogenes and tumor suppressor genes involved in cell proliferation, apoptosis, motility, and invasion. Many miRNAs have been found specifically associated with tumor type, tumor stage, and patient survival. Therefore, miRNAs are now being sincerely investigated as a source of potential biomarkers for the effective management of gastric cancer. Availability of such markers will also assist clinicians in designing precision medicine regimes for personalized treatment of the GC patients and provide potential targets for future drug development. This review summarizes the current knowledge about microRNA markers and their applicability in the diagnosis, prognosis, and prediction of treatment response in gastric cancer.

Keywords Gastric cancer · Biomarkers · MicroRNAs · Diagnosis · Prognosis · Targeted therapy

---

P. C. Sharma (✉) · R. Verma
University School of Biotechnology, Guru Gobind Singh Indraprastha University, New Delhi, India
e-mail: pcsharma@gmail.com; prof.pcsharma@gmail.com

P. Veera Bramhachari, N. R. R. Neelapu (eds.), *Recent Advancements in Biomarkers and Early Detection of Gastrointestinal Cancers*, Diagnostics and Therapeutic Advances in GI Malignancies, https://doi.org/10.1007/978-981-15-4431-6_6

6.1 Introduction

Cancer is the second leading cause of death globally accounting for 9.6 million deaths in 2018. Gastric cancer, a heterogeneous disease, is the sixth most common cancer with 1.03 million cases and the third most common cause of 7,83,000 cancer-related deaths. Although the rank of GC incidences has declined from fourth to sixth recently, the number of mortality cases has increased by 5.7%. Approximately, 70% of GC deaths occur in developing countries. Due to its asymptomatic behavior, GC is mostly diagnosed at an advanced stage. The available markers including the most known carcinoembryonic antigen (CEA) and carbohydrate antigen 19-9 (CA 19-9) lack consistency at early stages as compared to advanced stages. Therefore, the development of novel-sensitive biomarkers is imperative for early diagnosis of GC.

MicroRNAs (miRNAs) is a small class of non-coding RNAs of 20–24 base sequence that regulates gene expression at transcriptional and post-transcriptional level and plays a significant role in various physiological and pathological processes (Bartel 2004; Lee et al. 2003). miRNAs have been found to express aberrantly in cancer tissues. Apart from tissues, miRNAs can also be detected in serum, plasma, urine, tears, gastric juice as well. miRNAs traverse into biofluids through exosome particles or microvesicles that protect miRNAs from RNase degradation (Ma et al. 2013). Analysis of plasma and serum remains the most extensively used noninvasive method facilitating screening for miRNA-based diagnostic biomarkers (Link and Kupcinskas 2018). Role of miRNAs has been explored earnestly in oncogenesis, apoptosis, and tumor progression (Ekimler and Sahin 2014; Tian et al. 2014). miRNAs have also shown specific association with tumor type, tumor stage, *Helicobacter pylori* infection and patient survival. The length (~22 bp) and the stability of miRNAs under severe conditions including varying pH and temperatures give an advantage to evaluate them as biomarkers. Various studies have explored the role of miRNA in cancers, and it has been reported that China is the leading researcher in miRNA studies in GC followed by Japan, Taiwan, S. Korea, and Poland. There has been a remarkable increase in the number of miRNA-based studies in GC in this decade (Link and Kupcinskas 2018). In this chapter, we emphasized the role of miRNAs in gene regulation and their potential as diagnostic and prognostic markers in gastric cancer.

6.2 Molecular Classification of GC

Molecular classification of GC has been attempted by different groups, of which the following three recent molecular classifications of GC have been reported here:

1. Singapore Researchers
2. Asian Cancer Research Group (ACRG)
3. The Cancer Genome Atlas (TCGA)

A molecular classification of GC based on gene expression patterns made by researchers in 2013, grouped GC into the following three subtypes (Lei et al. 2013):

(a) Proliferative: This subtype displays high levels of genomic instability, *TP53* mutations, and DNA hypomethylation.
(b) Metabolic: Tumors of this subtype are associated with higher anaerobic glycolysis that makes the cells more sensitive to 5-fluorouracil therapy.
(c) Mesenchymal: Tumors of the mesenchymal subtype exhibit features of cancer stem cells and sensitivity to PIK3CA-AKT-mTOR pathway inhibitors.

Another molecular classification based on molecular alterations, disease progression, and prognosis proposed by the Asian Cancer Research Group (ACRG) in 2015 (Fig. 6.1) has categorized GC into four subtypes (Cristescu et al. 2015):

(a) Mesenchymal like type: It accounts for 15.3% of gastric tumors and includes tumors showing diffuse histology with the worst prognosis. They tend to occur at an advanced stage and an early age. It also showed a loss of CDH1 expression and the highest frequency (63%) of reoccurrence among the four subtypes. Microsatellite unstable tumors: This subtype represented by intestinal histology exhibits the best prognosis and the lowest frequency (22%) of reoccurrence among all subtypes of ACRG classification and predominantly arises at an early stage of GC.
(b) TP53 active: Tumors of this subtype are characterized by the presence of TP53 mutations, frequent EBV infection, and intermediate prognosis and reoccurrence rates.
(c) TP53 inactive: This subtype is marked by the absence of TP53 mutations, intermediate prognosis, and reoccurrence rates. Recurrent focal amplifications in RTKs had also been observed in the group.

One of the most recent and known classifications has been proposed by the Cancer Genome Atlas Group (TCGA) in 2014 on the basis of copy number variation (CNV), RNA sequencing, miRNA sequencing, exome sequencing, methylation status, and reverse phase protein assay (Cancer Genome Atlas Research Network 2014). This classification accommodates GC into the following four subtypes (Fig. 6.1):

(a) EBV-positive GC: This subtype represents moderately to poorly differentiated adenocarcinoma found in 9% of GC cases and is characterized by the association with Epstein–Barr virus, frequent PIK3CA mutations, and elevated expression of programmed death ligands 1 and 2 (PD-L1 and PD-L2). EBV-positive cancers are more prevalent in males (81% cases), particularly at young age and mainly located in fundus and body region of the stomach.
(b) Microsatellite unstable GC: This subtype characterized by microsatellite instability (MSI) is found in 22% of GC and has been associated with intestinal histology. MSI unstable GC shows CpG island methylation phenotype, including hypermethylation of the MLH1 promoter. Mutational analysis of MSI samples has identified 37 significantly mutated genes including TP53, PIK3A,

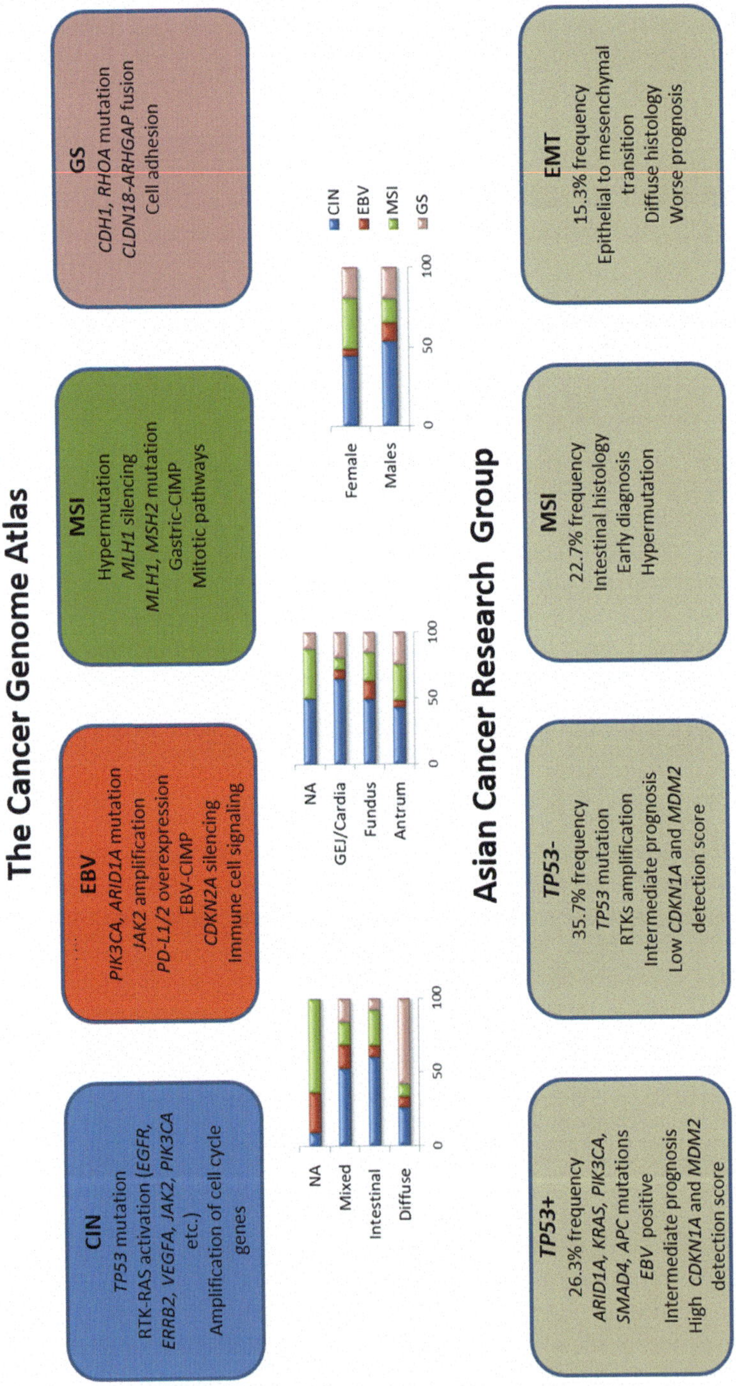

Fig. 6.1 Molecular classification of gastric cancer

KRAS, and ARID1A. Unlike colorectal cancer, BRAF and V600E mutations are not associated with microsatellite unstable GCs. It is more prevalent in females and found mainly in the antrum and pylorus regions. Alteration in MMR genes like *MLH*1 and *MSH2* leads to dysfunctioning of MMR system.

(c) GC with chromosomal instability: This subtype GC account for 50% of incidences that are located predominantly in the gastro-esophageal junction. Association of the intestinal type with copy number gains of chromosomes 8q, 17q, and 20q and diffuse type with gains of chromosomes 12q and 13q has been observed in GC with CIN. The chromosomal instability leads to the loss or gain of function of tumor suppressor and oncogenes. Mutation in TP53 gene, RTKs (receptor tyrosine kinases), and amplification of cell cycle genes are frequent in this subgroup. Amplification in oncogene pathways including MAPK signaling, RAS signaling is also an important feature.

(d) Genomically stable (GS) GC: The subtype is represented by 20% of GC incidences, diffuse histology, early age diagnosis, and comparable occurrence in males and females. Histologically, 25% tumors are located in the antrum, 20% in the gastro-esophageal junction and cardia, and approximately 15% in body and fundus. A recurrent interchromosomal translocation involving CLDN18 and ARHGAP26 has been found implicated in this subtype. The main somatic mutations observed in GS-GCs involve *CDH1*, *ARID1A*, and *RHOA* genes.

6.3 Role of miRNA in Gene Regulation

Aberrant miRNA expression has been found associated with tumorigenesis. Various studies have suggested that miRNAs play a crucial role in gene regulation. A schematic representation of upregulated and downregulated genes involved in GC is given in Fig. 6.2. Convincingly, miRNAs act as critical gene regulators involved in many biological processes.

MicroRNA-targeted tumor suppressor genes show a significantly reduced expression. miRNA-126 which regulates a tumor suppressor gene *PLK2* showed decreased expression in GC tissues. Moreover, miR-126 itself acts as tumor suppressor inhibiting GC cell invasion by targeting *Crk* gene. It also serves as an oncogene by targeting *SOX2* gene in GC (Liu et al. 2014). Inhibition of expression of other tumor suppressor genes, *PDCD4* and *PTEN*, by miRNA-21 results in growth, migration, and invasion of cancer cells in GC (Li et al. 2014).

Similarly, miR-124 suppresses the cell proliferation, migration, and invasion by targeting Rho-associated coiled-coil containing protein kinase 1 (*ROCK1*) (Hu et al. 2014). miR-148a gets inactivated by hypermethylation of the promoter region (Fujita et al. 2010). miRNA-148a suppresses tumor cell invasion by downregulating *ROCK1* (Zheng et al. 2011). Downregulation of miR-125a-5p targets *E2F3* and has been associated with GC metastasis. miR-106a, induced by SP1 and EGR1, downregulates the expression of *IL10* and acts as a regulatory element (Sharma et al. 2009).

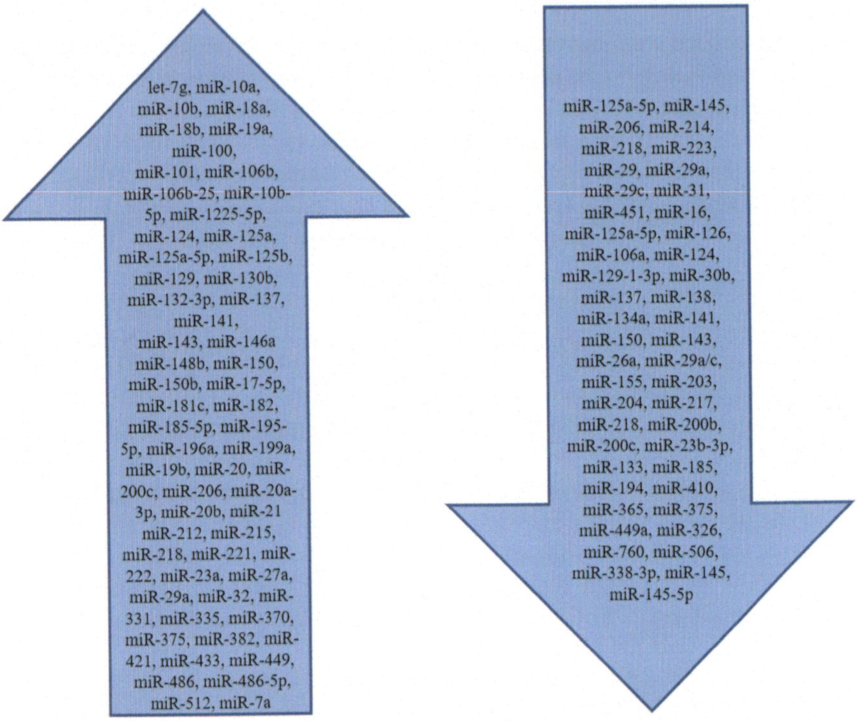

Fig. 6.2 Schematic diagram of differentially expressed miRNAs in gastric cancer

Increased DNA methylation of miR-210 has been linked with GC samples infected with *H. pylori.* Enhanced proliferation resulted from epigenetic silencing of miR-210 in gastric epithelial cells has been observed (Kiga et al. 2014). *H. pylori* infection led to the downregulation of miR-375 by targeting *JAK2* (Janus kinase 2) demonstrating that the JAK2-STAT3 pathway regulated by miR-375 is implied in *H. pylori* induced GC (Miao et al. 2014). miRNAs also play a role in angiogenesis. HIF-1α induced miR-382 targets tumor suppressor gene *PTEN* and acts as an oncogene promoting angiogenesis (Seok et al. 2014).

Expression of miRNA let-7 has been found to reduce the expression of *HRAS, KRAS,* and *NRAS* genes. *RAB40C*, a target gene of let-7, has been shown to play a significant role in gastric tumorigenesis (Yang et al. 2011). Enhancer of zeste homolog 2 (*EZH2*) contributes to the epigenetic silencing of target genes and regulates the survival and metastasis of cancer cells. The genomic loss of miR-101 in cancer leads to overexpression of *EZH2*, resulting in cancer progression. Overexpression of *EZH2* has been observed in aggressive solid tumors (Varambally et al. 2008).

6.4 Gastric Cancer Biomarkers

Screening for genomic biomarkers could lead to a better management of GC facilitating early diagnosis, prognosis, and predictable treatment response. Various new approaches have emerged which could be explored for the development of GC biomarkers.

Somatic alterations in short iterations of DNA sequences lead to genomic instability which may result in tumorigenesis. A random clinical trial reported the variation in prognosis of MSI-high and MSS/MSI-low gastro-esophageal cancer when treated with surgery alone and in combination with perioperative chemotherapy. The trial also indicated insignificance of perioperative chemotherapy in MSI-high cases (Smyth et al. 2017). 15–30% of gastric tumors showed MSI, particularly as a result of epigenetic silencing through promoter methylation of *MLH1* (Pinto et al. 2000). Microsatellite-positive tumors with *PIK3CA* mutations have been effectively treated with *PIK3CA* inhibitors as personalized therapy regime in GC patients (Zang et al. 2012). Instability at mononucleotide repeats in *CCDC150, CEP164, CNOT1, KIAA2018, MIS18BP1, RNPC3,* and *TGFBR2* has been reported in 63% of the MSI-positive GC samples (Yoon et al. 2013).

Modifications in the epigenome such as histone modifications and DNA methylation have been related to tumorigenesis in different cancers. Inactivation of tumor suppressor genes by methylation in the promoter region of the genes is a well-known feature observed in GC. Serum-based diagnostic markers (*CDH1, CHFR, P15. P16, RARβ, RUNA3*, etc.) exhibiting defective DNA methylation in GC has been previously presented (Qu et al. 2013). CHFR promoter methylation has also been linked to differentiation and lymph node status of GC (Ding et al. 2018). Loss of *FAT4* expression in methylated GC cell lines was observed by Yoshida and co-workers (Yoshida et al. 2017). Epigenetic profiles could serve as early diagnostic and prognostic biomarker in GC.

Experimental studies have demonstrated the implication of genetic polymorphism in Interleukin-1β in GC (Drici et al. 2016). Single nucleotide polymorphism in *CD44* gene has been suggested prognostic biomarker for early recurrence in GC (Suenaga et al. 2015). *CDH1, CSMD3, LRP1B, PIK3CA, ARID1A, TP53, SYNE1,* and *PKHD1* were among the top mutated genes displaying copy number variations in GC patients (Kuboki et al. 2016). Another study reported copy number variation in *KRAS*, *JAK2*, *CD274,* and *PDCD1LG2* (Hou et al. 2015).

6.5 miRNA Biomarkers

Gastric cancer being asymptomatic in nature is often diagnosed at an advanced stage. Thus, the need for biomarkers to detect GC at early stages is the primary objective of cancer management. Researchers have been exploring the feasibility of utilizing miRNAs as biomarkers considering the expression changes in tumor tissues and

Table 6.1 Important upregulated miRNAs involved in GC with their target genes and function

miRNA	Target gene(s)	Function
miR-92	*FXR*	Invasion, Proliferation
miR-21	*PTEN, TIMP1*	Apoptosis, Invasion, Migration, Proliferation
miR-107	*DICER1*	Invasion, Migration
miR-25	*FBXW7*	Invasion, Migration, Proliferation
miR-106b	*PTEN*	Invasion, Migration
miR-500	*NF-kB*	Apoptosis, Proliferation
miR-124	*ROCK1*	Invasion, Proliferation
miR-146a	*EGFR*	Invasion, Migration
miR-150	*EGR2*	Apoptosis, Proliferation
miR-200c	*CDH, RHO*	Metastasis
miR-210	*STMN1, DIMT1*	Angiogenesis
miR-181a	*PTEN*	Proliferation
miR-181c	*KRAS, NOTCH4*	Proliferation
miR-183	*PTEN*	Migration
miR-449	*MET, SIRT1, CDK6*	Apoptosis, Cell cycle, Proliferation
miR-221	*CDKN1A, CDKN1B, CDKN1C*	Cell cycle
miR-222	*CDKN1A, CDKN1B, CDKN1C*	Cell cycle
miR-421	*BAX, BCL-2*	Oncogenes
miR-362	*NF-kB*	Anti-apoptotic
miR-382	*PTEN*	Angiogenesis
miR-377	*P53, PTEN, TIMP1*	Proliferation
miR-520d-3p	*EPHA2*	Inhibits proliferation and invasion
miR-508	*INPP5J*	Invasion, Migration, Proliferation
miR-942	*SFRP4, GSK3B, TLE1*	Proliferation
miR-1288	*FOXO1*	Proliferation
miR-125a-5p	*ERBB2, E2F3*	Invasion, Metastasis, Migration, Proliferation

biofluids as well indicating their involvement in proliferation, invasion, metastasis, and tumorigenesis (Tables 6.1 and 6.2).

6.5.1 Diagnostic Markers

Various studies based on the expression profile and next-generation sequencing has provided useful evidence to highlight the diagnostic potential of miRNA in GC.

6.5.1.1 Blood-Based Markers

mi-375 showed decreased expression in distal gastric adenocarcinoma tissues and significant downregulation in serum samples in comparison to control samples

Table 6.2 Important downregulated miRNAs involved in GC with their target genes and function

miRNA	Target gene(s)	Function
miR-16	*P53*	Proliferation
miR-125a-5p	*ERBB2, E2F3*	Invasion, Metastasis, Proliferation
miR-126	*PI3KR2, CrK, PLK2*	Invasion, Metastasis, Proliferation
miR-106a	*EGFL7, E2F1*	Invasion, Migration
miR-124	*ROCK1*	Inhibits proliferation
miR-129-1-3p	*BDKRB2, PDCD2*	Inhibits migration
miR-30b	*PAI-1*	Apoptosis
miR-137	*AKT2*	Proliferation
miR-138	*NF-Kb*	Proliferation
miR-134a	*FSCN, MMP14*	Invasion, Migration
miR-141	*ZEB1, ZEB2*	Invasion, Migration
miR-150	*ZEB1*	EMT
miR-143	*TLR2*	Invasion, Migration
miR-26a	*FGF9*	Metastasis, Proliferation
miR-29a/c	*VEGF*	Metastasis, Proliferation
miR-155	*c-myc*	Invasion, Proliferation
miR-203	*E-cadherin*	EMT, Migration
miR-204	*SOX4*	Invasion, Proliferation
miR-217	*EZH2*	Invasion, Metastasis, Proliferation
miR-218	*ROBO1*	Apoptosis, Invasion, Proliferation
miR-200b	*DNMT3A, DNMT3B, SP1*	Proliferation
miR-200c	*ZEB1, ZEB2*	Invasion, Migration
miR-23b-3p	*ATG12, HMGB2*	Chemoresistance
miR-133	*CDC42-PAK*	Invasion, Migration, Proliferation
miR-185	*DNMT1, CDC42*	Metastasis
miR-194	*RBX1*	Migration, Proliferation
miR-410	*MDM2*	Inhibits invasion and migration
miR-365	*Cyclin D1, BCL-2*	Apoptosis
miR-375	*PDK1, JAK2*	Inhibits proliferation
miR-449a	*CDK6*	Apoptosis
miR-326	*FSCN1*	Migration, Proliferation
miR-760	*HIST1H3D*	Migration
miR-506	*YAP-1*	Invasion, Proliferation
miR-338-3p	*SMO*	Apoptosis
miR-145	*ETS1*	Angiogenesis, Invasion, Migration
miR-145-5p	*TLR4, KLF5*	Inhibits proliferation

($p < 0.001$) (Tsujiura et al. 2010). Tsujiura et al. reported increased expression levels of miR-21 ($p = 0.05$), miR-17-5p ($p = 0.006$), miR-106a ($p = 0.008$), and miR-106b ($p < 0.001$) in plasma. Decreased expression of let-7a ($p = 0.002$) was also reported suggesting the role of all these miRNAs as tumor markers for GC diagnosis (Tsujiura et al. 2010). *H. pylori* infection has been linked with miRNA expression levels. It was proposed that serum miR-106b was significantly overexpressed before and after eradication of *H. pylori* as compared to healthy controls where miR-21 showed significantly high expression after *H. pylori* eradication when compared with healthy controls in GC patients (Shiotani et al. 2013).

The plasma levels of miR-223 ($p < 0.001$) and miR-21 ($p < 0.001$) were found significantly higher in GC patients than in healthy controls while miR-218 was significantly lower ($p < 0.001$). The combined ROC analysis of all the three miRNA revealed AUC value of 0.953 in discriminating GC patients from healthy controls. Also, a correlation between expression levels of miR-223 with *H. pylori* infection was reported (Li et al. 2012).

Expression of three miRNAs has been validated for early detection of GC using qRT-PCR. Here, the level of expression was first checked in a cohort of 30 patients and then validated on a sample size of 60 patients diagnosed with GC. Upregulation of miR-106b, miR-20a, and miR-221 ($p < 0.05$) in plasma suggested their potential role as early-stage biomarker. The area under ROC curves was 0.773 for miR-106b, 0.859 for miR-20a, and 0.796 for miR-221. The three markers might be useful together as a panel of biomarkers for diagnosis (Cai et al. 2013). Liu et al. reported elevated levels of miR-187 ($p = 0.0016$), miR-371-5p ($p < 0.0009$), and miR-378 ($p < 0.0001$) in serum samples of GC patients. The ROC curve area of miR-378 was 0.861 with 87.5% sensitivity and 70.73% specificity. Further, the inclusion of miR-187 and miR-371-5p did not improve the discrimination value significantly (Liu et al. 2012).

Plasma samples of 12 GC patients with distant metastasis observed significantly lower and higher levels of miR-122 and miR-192 with AUC 0.808 and 0.732, respectively (Chen et al. 2014a). Zhu et al. screened 36 patients diagnosed with gastric cardia adenocarcinoma (GNCA) along with 160 cancer-free controls for recording the expression level of miRNA. The study revealed overexpression of miR-16, miR-25, miR-92a, miR-451, and miR-486-5p as a suggestive biomarker in detecting the early stage GC (Zhu et al. 2014).

A microarray experiment was performed on 123 patients and 111 healthy controls to identify deregulated miRNAs in GC. Overexpression along with high sensitivity (86.7%) and specificity (85.5%) of miR-627, miR-629, and miR-652 were observed which show their potential for use as a panel of potential biomarkers (Shin et al. 2015). Overexpression of miRNA-185, miR-20a, miR-210, miR-25, miR-92b ($p < 0.05$), miR-10b-5p, miR-132-3p, miR-185-5p, miR-195-5p, miR-20a-3p, and miR-296-5p has been observed by different group of researchers (Zhou et al. 2015; Huang et al. 2017). Involvement of miR-940 in the initiation and progression of GC through NF-κB and Wnt/β signaling pathway was predicted in plasma and cell lines of GC patients (Liu et al. 2016).

6.5.1.2 Tissue-Based Markers

The expression levels of miR-106a, miR-421, and miR-21 were significantly higher while the level of miR-31 significantly downregulated in GC tissue samples (Xiao et al. 2009; Chan et al. 2008; Jiang et al. 2010; Zhang et al. 2010). miRNA with dysregulated expression can play a tumor suppressor or an oncogenic role. Overexpressed miR-21 binds to *PDCD4*, a tumor suppressor gene, and inhibits its protein expression. The miR-21 expression has been related to tumor size, depth of invasion, lymph node metastasis, and vascular invasion (Li et al. 2012; Chan et al. 2008; Motoyama et al. 2010). Other miRNAs such as miR-32, miR-182, miR-143, and miR-106a have been found upregulated in GC tissues (Xiao et al. 2009; Li et al. 2011). The expression level of miR-106a is closely related to the size of the tumor, differentiation status, lymph node involved, and distant metastasis (Xiao et al. 2009). miR-31, miR-218, and miR-223 are tissue-based downregulated miRNA biomarkers in GC (Li et al. 2012; Zhang et al. 2010). The sensitivity of miR-421 in GC tissues has been observed to be more than serum carcinoembryogenic antigen which indicates its potential as a diagnostic marker (Jiang et al. 2010).

6.5.1.3 Biofluid-Based Markers

Gastric juice: Although the collection of gastric juice from the patients through gastroscopy or evacuated tubes is invasive but the examination of miRNA in gastric juice could result in better treatment prediction of GC. miRNAs have been observed to withstand low (pH = 1) to high (pH = 13) making them suitable for gastric juice-based studies (Chen et al. 2008). Discrimination of GC from healthy and benign gastric disease with miR-421 and miR-133a in gastric juice has been realized (Shao et al. 2016; Zhang et al. 2012). The miR-21 and miR-106a in gastric juice, when used together, have been observed to detect GC up to 98% (Cui et al. 2013). The low expression levels of miR-129-1-3p and miR-129-2-3p in gastric juice were analyzed in GC patients (Cui et al. 2013).

Urine: Diagnostic value of miR-376c has been observed in GC patients where the level of its expression in urine was found to be increased (Hung et al. 2017). Another study exhibited high expression of miR-21-5p in urine samples of GC patients compared to the healthy controls. The levels of miR-21-5p significantly reduced after surgical resection (Kao et al. 2017).

Exosomes: They are small vesicles enclosed by lipid bilayer membrane in the extracellular environment that are secreted by cells and contain a variety of molecules including miRNAs. Alike miRNA in gastric juice is protected from varying pH; the miRNA is protected in exosomes from ribonuclease degradation (Valadi et al. 2007). Levels of miR-221 in exosomes from peripheral blood were found to be increased by 2.5 fold in GC patients compared to healthy controls. Exosomal miR-106a-5p and miR-19b-3p were found to be elevated in GC patients and exhibited 81% detection ability when combined together (Wang et al. 2017).

6.5.2 Prognostic Markers

Apart from being suitable diagnostic markers, the prospective role of miR-21, miR-106a, and miR-106b as prognostic markers was reported in plasma samples of GC patients. Overexpression of miR-21 has been correlated with vascular invasion ($p = 0.0311$) and could be used as an independent prognostic biomarker in GC. Correlation of tumor size and stage with miR-21 expression has been established (Komatsu et al. 2013; Kim et al. 2013).

miR-20a and miR-17-5p have been found significantly correlated with differentiation, staging, and poor overall survival. A decrease in the expression level of miR-17-5p and miR-20a was observed in response to chemotherapy (Wang et al. 2012). miR-20a solely showed the potential ability to be a prognostic marker. Expression of miR-17-5p as a prognostic marker and in the assessment chemotherapeutic effects on GC has been detected (Komatsu et al. 2013; Wang et al. 2012).

The expression level of miR-196a in serum and tissue of GC patients was found correlated with progression and relapse of GC (Tsai et al. 2012). Another microRNA, miRNA-195-5p, with prognostic value has been observed (Gorur et al. 2013). Low expression of let-7a miRNA in serum and tissue samples of GC was observed and correlated with lymph node metastasis, depth of invasion, staging, tumor size, and progression of GC (Wang et al. 2013). Association of expression of miRNA with GC metastasis has been analyzed. Decreased expression of miR-218 in the serum sample of GC patients and high expression of mi-214 in plasma and serum samples has been associated with GC metastasis (Xin et al. 2014; Zhang et al. 2015). The levels of miR-214 were significantly decreased after surgical resection.

Overexpression of miR-25 is correlated with lymph node metastasis by targeting TOB1 (Li et al. 2015). Correlation between low levels of miR-203 and metastasis was found and an inverse relation between GC development and level of miR-203 was analyzed by Imaoka group (Imaoka et al. 2016). miR-29 and miR-106b are tissue-based miRNAs found associated with poor prognosis with low and high expression levels, respectively. miR-125a-5p and miR-206 are independent prognostic factor showing downregulation in GC patients (Yang et al. 2013; Nishida et al. 2011). A study analyzed seven miRNAs (miR-10b, miR-21, miR-223, miR-338, let-7a, miR-30a-5p, miR-126) significantly related to recurrence-free periods and overall survival of patients (Li et al. 2010). Upregulated miR-125b, miR-199a, and miR-100 have shown association with progression of GC (Ueda et al. 2010).

6.5.3 Therapeutic Markers

The potential prognostic and diagnostic markers undergo clinical trials to improve cancer treatment regimes. Evidence suggests that miRNA therapeutics have been evaluated in both preclinical and clinical settings. Moreover, the probability of miRNA in drug resistance has been explored. The expression of miR-218 has been demonstrated to increase in vitro cell chemosensitivity to drug cisplatin and decrease tumor growth (Zhang et al. 2014).

Overexpression of miR-362 induced cell proliferation and resistance to cisplatin induced apoptosis in BGC-823 and SGC-7901 GC cells (Xia et al. 2014). miR-129 has been used as a novel therapeutic target in gastrointestinal marker (Fesler et al. 2014). An inverse relation between the expression level of miR-196a/miR-196b and *RDX* protein levels have been observed. Reduced miR-196a/miR-196b levels or increased level of the *RDX* gene have a potential therapeutic role in GC metastasis. Cisplatin resistance of GC cell lines is found to be regulated by miR-503 by targeting *IGF1R* and *BCL2* (Wang et al. 2014). Similarly, miR-1271 targets BCL2, IGF1R, IRS1, and mTOR genes and has shown to regulate cisplatin resistance in GC cell lines (Yang et al. 2014).

miRNAs such as miR-92b and miR-422a have been found associated with relapse following chemotherapy in GC (Omura et al. 2014). Shen et al. described the importance of clinical efficacy of DNA damage inducing chemotherapeutic drug by reducing drug resistance. The study analyzed that doxorubicin downregulates HDAC1 protein expression which is a target gene of miR-520 h (Shen et al. 2014). miR-1207-5p and miR-1266 are significantly decreased in GC tissues and their ectopic expression inhibits tumor growth by suppressing hTERT. These miRNA provides a novel therapeutic approach for GC treatment (Chen et al. 2014b).

6.6 Future Perspectives

miRNA have emerged as crucial translational gene regulators in cancers including gastric cancer. The discovery of noninvasive and specific biomarkers which could provide early detectability and personalized treatment is needed. Although the development of miRNA-related biomarkers is still in the preclinical stages, they hold huge potential as biomarkers facilitating early diagnosis, prognosis, and therapeutics in gastric cancer. Being a heterogeneous disease, GC shows different outcomes in the similar clinical and pathological conditions. Therefore, the novel biomarkers need to be based on genome analysis ensuring prevention and treatment of the disease. Molecular classification in combination with the histological classification of GC could be used as a platform to explore the underlying mutations in GC and to design prognostic and therapeutic regimes. Several reports in contemporary literature have advocated the use of single/combinations of biomarkers in GC that can predict favorable or unfavorable response towards single/multidrug treatment regimes (Duraes et al. 2014). MicroRNAs with multi-functional characteristics, i.e., a single microRNA with diagnostic, prognostic, and therapeutic role are desired for the development of efficient biomarkers (Fig. 6.3). GC-specific miRNA have been associated with tumor formation, proliferation, and metastasis. Future studies with identification and validation of miRNA-based diagnostic, prognostic, and therapeutic biomarker will aid to the better understanding and management of GC.

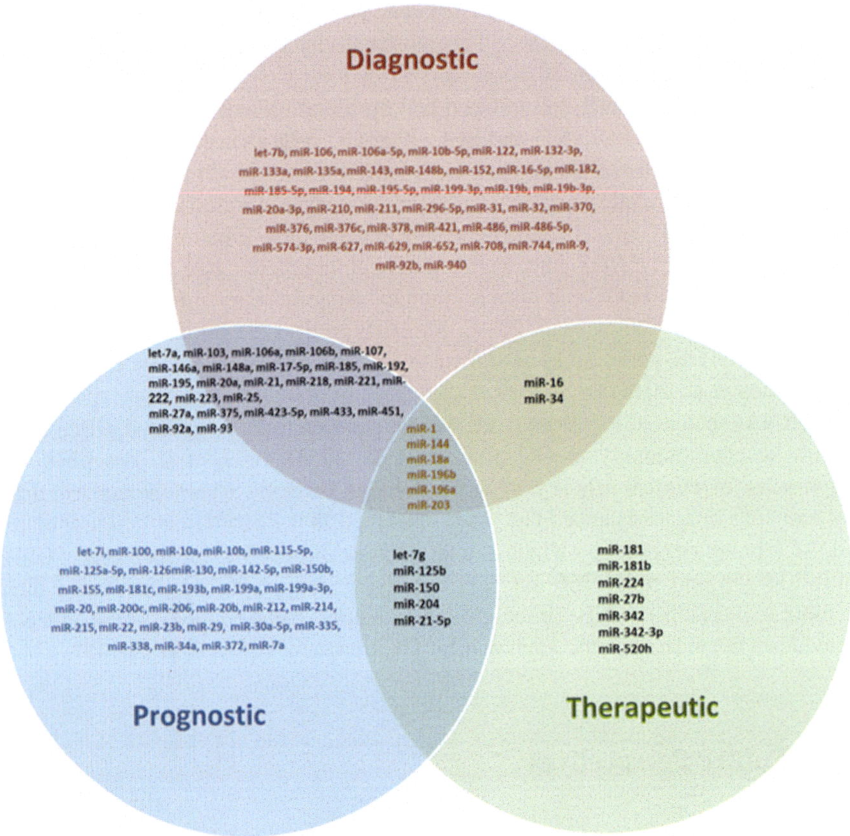

Fig. 6.3 Multifunctional role of miRNA biomarkers in gastric cancer

6.7 Conclusions

We have presented the role of miRNA in GC and their potential use as future diagnostic, prognostic, and therapeutic biomarkers. Apart from current conventional tumor antigens such as CEA, CA19.9, and CA72.4, there is an urgent and strong need for the development of novel biomarkers, single or in combination, with high sensitivity and specificity for the screening of GC. These miRNA-based biomarkers should further be explored exhaustively for clinical testing to facilitate the diagnosis, prognosis, and personalized treatment of the disease.

References

Bartel DP (2004) MicroRNAs: genomics, biogenesis, mechanism, and function. Cell 116 (2):281–297

Cai H, Yuan Y, Hao YF, Guo TK, Wei X, Zhang YM (2013) Plasma microRNAs serve as novel potential biomarkers for early detection of gastric cancer. Med Oncol 30(1):452

Cancer Genome Atlas Research Network (2014) Comprehensive molecular characterization of gastric adenocarcinoma. Nature 513(7517):202–209

Chan SH, Wu CW, Li AF, Chi CW, Lin WC (2008) miR-21 microRNA expression in human gastric carcinomas and its clinical association. Anticancer Res 28(2A):907–911

Chen X, Ba Y, Ma L, Cai X, Yin Y, Wang K, Guo J, Zhang Y, Chen J, Guo X, Li Q, Li X, Wang W, Zhang Y, Wang J, Jiang X, Xiang Y, Xu C, Zheng P, Zhang J, Li R, Zhang H, Shang X, Gong T, Ning G, Wang J, Zen K, Zhang J, Zhang CY (2008) Characterization of microRNAs in serum: a novel class of biomarkers for diagnosis of cancer and other diseases. Cell Res 18(10):997–1006

Chen Q, Ge X, Zhang Y, Xia H, Yuan D, Tang Q, Chen L, Pang X, Leng W, Bi F (2014a) Plasma miR-122 and miR-192 as potential novel biomarkers for the early detection of distant metastasis of gastric cancer. Oncol Rep 31(4):1863–1870

Chen L, Lu MH, Zhang D, Hao NB, Fan YH, Wu YY, Wang SM, Xie R, Fang DC, Zhang H, Hu CJ, Yang SM (2014b) miR-1207-5p and miR-1266 suppress gastric cancer growth and invasion by targeting telomerase reverse transcriptase. Cell Death Dis 5:e1034

Cristescu R, Lee J, Nebozhyn M, Kim KM, Ting JC, Wong SS, Liu J, Yue YG, Wang J, Yu K, Ye XS, Do IG, Liu S, Gong L, Fu J, Jin JG, Choi MG, Sohn TS, Lee JH, Bae JM, Kim ST, Park SH, Sohn I, Jung SH, Tan P, Chen R, Hardwick J, Kang WK, Ayers M, Hongyue D, Reinhard C, Loboda A, Kim S, Aggarwal A (2015) Molecular analysis of gastric cancer identifies subtypes associated with distinct clinical outcomes. Nat Med 21(5):449–456

Cui L, Zhang X, Ye G, Zheng T, Song H, Deng H, Xiao B, Xia T, Yu X, Le Y, Guo J (2013) Gastric juice microRNAs as potential biomarkers for the screening of gastric cancer. Cancer 119 (9):1618–1626

Ding Y, Lian HF, Du Y (2018) Clinicopathological significance of CHFR promoter methylation in gastric cancer: a meta-analysis. Oncotarget 9(11):10083–10090

Drici Ael-M, Moulessehoul S, Tifrit A, Diaf M, Turki DK, Bachir M, Tou A (2016) Effect of IL-1beta and IL-1RN polymorphisms in carcinogenesis of the gastric mucosa in patients infected with Helicobacter pylori in Algeria. Libyan J Med 11:31576

Duraes C, Almeida GM, Seruca R, Oliveira C, Carneiro F (2014) Biomarkers for gastric cancer: prognostic, predictive or targets of therapy? Virchows Arch 464(3):367–378

Ekimler S, Sahin K (2014) Computational methods for microRNA target prediction. Genes (Basel) 5(3):671–683

Fesler A, Zhai H, Ju J (2014) miR-129 as a novel therapeutic target and biomarker in gastrointestinal cancer. Onco Targets Ther 7:1481–1485

Fujita Y, Kojima K, Ohhashi R, Hamada N, Nozawa Y, Kitamoto A, Sato A, Kondo S, Kojima T, Deguchi T, Ito M (2010) MiR-148a attenuates paclitaxel resistance of hormone-refractory, drug-resistant prostate cancer PC3 cells by regulating MSK1 expression. J Biol Chem 285 (25):19076–19084

Gorur A, Balci Fidanci S, Dogruer Unal N, Ayaz L, Akbayir S, Yildirim Yaroglu H, Dirlik M, Serin MS, Tamer L (2013) Determination of plasma microRNA for early detection of gastric cancer. Mol Biol Rep 40(3):2091–2096

Hou Y, Wu K, Shi X, Li F, Song L, Wu H, Dean M, Li G, Tsang S, Jiang R, Zhang X, Li B, Liu G, Bedekar N, Lu N, Xie G, Liang H, Chang L, Wang T, Chen J, Li Y, Zhang X, Yang H, Xu X, Wang L, Wang J (2015) Comparison of variations detection between whole-genome amplification methods used in single-cell resequencing. Gigascience 4:37

Hu CB, Li QL, Hu JF, Zhang Q, Xie JP, Deng L (2014) miR-124 inhibits growth and invasion of gastric cancer by targeting ROCK1. Asian Pac J Cancer Prev 15(16):6543–6546

Huang Z, Zhu D, Wu L, He M, Zhou X, Zhang L, Zhang H, Wang W, Zhu J, Cheng W, Chen Y, Fan Y, Qi L, Yin Y, Zhu W, Shu Y, Liu P (2017) Six serum-based miRNAs as potential diagnostic biomarkers for gastric cancer. Cancer Epidemiol Biomark Prev 26(2):188–196

Hung PS, Chen CY, Chen WT, Kuo CY, Fang WL, Huang KH, Chiu PC, Lo SS (2017) miR-376c promotes carcinogenesis and serves as a plasma marker for gastric carcinoma. PLoS One 12(5): e0177346

Imaoka H, Toiyama Y, Okigami M, Yasuda H, Saigusa S, Ohi M, Tanaka K, Inoue Y, Mohri Y, Kusunoki M (2016) Circulating microRNA-203 predicts metastases, early recurrence, and poor prognosis in human gastric cancer. Gastric Cancer 19(3):744–753

Jiang Z, Guo J, Xiao B, Miao Y, Huang R, Li D, Zhang Y (2010) Increased expression of miR-421 in human gastric carcinoma and its clinical association. J Gastroenterol 45(1):17–23

Kao HW, Pan CY, Lai CH, Wu CW, Fang WL, Huang KH, Lin WC (2017) Urine miR-21-5p as a potential non-invasive biomarker for gastric cancer. Oncotarget 8(34):56389–56397

Kiga K, Mimuro H, Suzuki M, Shinozaki-Ushiku A, Kobayashi T, Sanada T, Kim M, Ogawa M, Iwasaki YW, Kayo H, Fukuda-Yuzawa Y, Yashiro M, Fukayama M, Fukao T, Sasakawa C (2014) Epigenetic silencing of miR-210 increases the proliferation of gastric epithelium during chronic Helicobacter pylori infection. Nat Commun 5:4497

Kim SY, Jeon TY, Choi CI, Kim DH, Kim DH, Kim GH, Ryu DY, Lee BE, Kim HH (2013) Validation of circulating miRNA biomarkers for predicting lymph node metastasis in gastric cancer. J Mol Diagn 15(5):661–669

Komatsu S, Ichikawa D, Tsujiura M, Konishi H, Takeshita H, Nagata H, Kawaguchi T, Hirajima S, Arita T, Shiozaki A, Kubota T, Fujiwara H, Okamoto K, Otsuji E (2013) Prognostic impact of circulating miR-21 in the plasma of patients with gastric carcinoma. Anticancer Res 33 (1):271–276

Kuboki Y, Yamashita S, Niwa T, Ushijima T, Nagatsuma A, Kuwata T, Yoshino T, Doi T, Ochiai A, Ohtsu A (2016) Comprehensive analyses using next-generation sequencing and immunohistochemistry enable precise treatment in advanced gastric cancer. Ann Oncol 27 (1):127–133

Lee Y, Ahn C, Han J, Choi H, Kim J, Yim J, Lee J, Provost P, Radmark O, Kim S, Kim VN (2003) The nuclear RNase III Drosha initiates microRNA processing. Nature 425(6956):415–419

Lei Z, Tan IB, Das K, Deng N, Zouridis H, Pattison S, Chua C, Feng Z, Guan YK, Ooi CH, Ivanova T, Zhang S, Lee M, Wu J, Ngo A, Manesh S, Tan E, The BT, So JB, Goh LK, Boussioutas A, Lim TK, Flotow H, Tan P, Rozen SG (2013) Identification of molecular subtypes of gastric cancer with different responses to PI3-kinase inhibitors and 5-fluorouracil. Gastroenterology 145(3):554–565

Li X, Zhang Y, Zhang Y, Ding J, Wu K, Fan D (2010) Survival prediction of gastric cancer by a seven-microRNA signature. Gut 59(5):579–585

Li X, Luo F, Li Q, Xu M, Feng D, Zhang G, Wu W (2011) Identification of new aberrantly expressed miRNAs in intestinal-type gastric cancer and its clinical significance. Oncol Rep 26 (6):1431–1439

Li BS, Zhao YL, Guo G, Li W, Zhu ED, Luo X, Mao XH, Zou QM, Yu PW, Zuo QF, Li N, Tang B, Liu KY, Xiao B (2012) Plasma microRNAs, miR-223, miR-21 and miR-218, as novel potential biomarkers for gastric cancer detection. PLoS One 7(7):e41629

Li L, Zhou L, Li Y, Lin S, Tomuleasa C (2014) MicroRNA-21 stimulates gastric cancer growth and invasion by inhibiting the tumor suppressor effects of programmed cell death protein 4 and phosphatase and tensin homolog. J BUON 19(1):228–236

Li BS, Zuo QF, Zhao YL, Xiao B, Zhuang Y, Mao XH, Wu C, Yang SM, Zeng H, Zou QM, Guo G (2015) MicroRNA-25 promotes gastric cancer migration, invasion and proliferation by directly targeting transducer of ERBB2, 1 and correlates with poor survival. Oncogene 34 (20):2556–2565

Link A, Kupcinskas J (2018) MicroRNAs as non-invasive diagnostic biomarkers for gastric cancer: current insights and future perspectives. World J Gastroenterol 24(30):3313–3329

Liu H, Zhu L, Liu B, Yang L, Meng X, Zhang W, Ma Y, Xiao H (2012) Genome-wide microRNA profiles identify miR-378 as a serum biomarker for early detection of gastric cancer. Cancer Lett 316(2):196–203

Liu LY, Wang W, Zhao LY, Guo B, Yang J, Zhao XG, Hou N, Ni L, Wang AY, Song TS, Huang C, Xu JR (2014) Mir-126 inhibits growth of SGC-7901 cells by synergistically targeting the oncogenes PI3KR2 and Crk, and the tumor suppressor PLK2. Int J Oncol 45(3):1257–1265

Liu X, Kwong A, Sihoe A, Chu KM (2016) Plasma miR-940 may serve as a novel biomarker for gastric cancer. Tumour Biol 37(3):3589–3597

Ma Y, Wang X, Jin H (2013) Methylated DNA and microRNA in body fluids as biomarkers for cancer detection. Int J Mol Sci 14(5):10307–10331

Miao L, Liu K, Xie M, Xing Y, Xi T (2014) miR-375 inhibits Helicobacter pylori-induced gastric carcinogenesis by blocking JAK2-STAT3 signaling. Cancer Immunol Immunother 63 (7):699–711

Motoyama K, Inoue H, Mimori K, Tanaka F, Kojima K, Uetake H, Sugihara K, Mori M (2010) Clinicopathological and prognostic significance of PDCD4 and microRNA-21 in human gastric cancer. Int J Oncol 36(5):1089–1095

Nishida N, Mimori K, Fabbri M, Yokobori T, Sudo T, Tanaka F, Shibata K, Ishii H, Doki Y, Mori M (2011) MicroRNA-125a-5p is an independent prognostic factor in gastric cancer and inhibits the proliferation of human gastric cancer cells in combination with trastuzumab. Clin Cancer Res 17(9):2725–2733

Omura T, Shimada Y, Nagata T, Okumura T, Fukuoka J, Yamagishi F, Tajika S, Nakajima S, Kawabe A, Tsukada K (2014) Relapse-associated microRNA in gastric cancer patients after S-1 adjuvant chemotherapy. Oncol Rep 31(2):613–618

Pinto M, Oliveira C, Machado JC, Cirnes L, Tavares J, Carneiro F, Hamelin R, Hofstra R, Seruca R, Sobrinho-Simoes M (2000) MSI-L gastric carcinomas share the hMLH1 methylation status of MSI-H carcinomas but not their clinicopathological profile. Lab Invest 80(12):1915–1923

Qu Y, Dang S, Hou P (2013) Gene methylation in gastric cancer. Clin Chim Acta 424:53–65

Seok JK, Lee SH, Kim MJ, Lee YM (2014) MicroRNA-382 induced by HIF-1alpha is an angiogenic miR targeting the tumor suppressor phosphatase and tensin homolog. Nucleic Acids Res 42(12):8062–8072

Shao J, Fang PH, He B, Guo LL, Shi MY, Zhu Y, Bo P, Zhen-Wen ZW (2016) Downregulated MicroRNA-133a in gastric juice as a clinicopathological biomarker for gastric cancer screening. Asian Pac J Cancer Prev 17(5):2719–2722

Sharma A, Kumar M, Aich J, Hariharan M, Brahmachari SK, Agrawal A, Ghosh B (2009) Posttranscriptional regulation of interleukin-10 expression by hsa-miR-106a. Proc Natl Acad Sci U S A 106(14):5761–5766

Shen Q, Yao Q, Sun J, Feng L, Lu H, Ma Y, Liu L, Wang F, Li J, Yue Y, Jin H, Wang X (2014) Downregulation of histone deacetylase 1 by microRNA-520h contributes to the chemotherapeutic effect of doxorubicin. FEBS Lett 588(1):184–191

Shin VY, Ng EK, Chan VW, Kwong A, Chu KM (2015) A three-miRNA signature as promising non-invasive diagnostic marker for gastric cancer. Mol Cancer 14:202

Shiotani A, Murao T, Kimura Y, Matsumoto H, Kamada T, Kusunoki H, Inoue K, Uedo N, Iishi H, Haruma K (2013) Identification of serum miRNAs as novel non-invasive biomarkers for detection of high risk for early gastric cancer. Br J Cancer 109(9):2323–2330

Smyth EC, Wotherspoon A, Peckitt C, Gonzalez D, Hulkki-Wilson S, Eltahir Z, Fassan M, Rugge M, Valeri N, Okines A, Hewish M, Allum W, Stenning S, Nankivell M, Langley R, Cunningham D (2017) Mismatch repair deficiency, microsatellite instability, and survival: an exploratory analysis of the Medical Research Council adjuvant gastric infusional chemotherapy (MAGIC) trial. JAMA Oncol 3(9):1197–1203

Suenaga M, Yamada S, Fuchs BC, Fujii T, Kanda M, Tanaka C, Kobayashi D, Fujiwara M, Tanabe KK, Kodera Y (2015) CD44 single nucleotide polymorphism and isoform switching may predict gastric cancer recurrence. J Surg Oncol 112(6):622–628

Tian SB, Yu JC, Kang WM, Ma ZQ, Ye X, Cao ZJ (2014) [MicroRNA and gastric cancer]. Zhongguo Yi Xue Ke Xue Yuan Xue Bao 36(2):214–217

Tsai KW, Liao YL, Wu CW, Hu LY, Li SC, Chan WC, Ho MR, Lai CH, Kao HW, Fang WL, Huang KH, Lin WC (2012) Aberrant expression of miR-196a in gastric cancers and correlation with recurrence. Genes Chromosomes Cancer 51(4):394–401

Tsujiura M, Ichikawa D, Komatsu S, Shiozaki A, Takeshita H, Kosuga T, Konishi H, Morimura R, Deguchi K, Fujiwara H, Okamoto K, Otsuji E (2010) Circulating microRNAs in plasma of patients with gastric cancers. Br J Cancer 102(7):1174–1179

Ueda T, Volinia S, Okumura H, Shimizu M, Taccioli C, Rossi S, Alder H, Liu CG, Oue N, Yasui W, Yoshida K, Sasaki H, Nomura S, Seto Y, Kaminishi M, Calin GA, Croce CM (2010) Relation between microRNA expression and progression and prognosis of gastric cancer: a microRNA expression analysis. Lancet Oncol 11(2):136–146

Valadi H, Ekstrom K, Bossios A, Sjostrand M, Lee JJ, Lotvall JO (2007) Exosome-mediated transfer of mRNAs and microRNAs is a novel mechanism of genetic exchange between cells. Nat Cell Biol 9(6):654–659

Varambally S, Cao Q, Mani RS, Shankar S, Wang X, Ateeq B, Laxman B, Cao X, Jing X, Ramnarayanan K, Brenner JC, Yu J, Kim JH, Han B, Tan P, Kumar-Sinha C, Lonigro RJ, Palanisamy N, Maher CA, Chinnaiyan AM (2008) Genomic loss of microRNA-101 leads to overexpression of histone methyltransferase EZH2 in cancer. Science 322(5908):1695–1699

Wang M, Gu H, Wang S, Qian H, Zhu W, Zhang L, Zhao C, Tao Y, Xu W (2012) Circulating miR-17-5p and miR-20a: molecular markers for gastric cancer. Mol Med Rep 5(6):1514–1520

Wang WN, Chen Y, Zhang YD, Hu TH (2013) The regulatory mechanism of CCR7 gene expression and its involvement in the metastasis and progression of gastric cancer. Tumour Biol 34(3):1865–1871

Wang T, Ge G, Ding Y, Zhou X, Huang Z, Zhu W, Shu Y, Liu P (2014) MiR-503 regulates cisplatin resistance of human gastric cancer cell lines by targeting IGF1R and BCL2. Chin Med J 127 (12):2357–2362

Wang N, Wang L, Yang Y, Gong L, Xiao B, Liu X (2017) A serum exosomal microRNA panel as a potential biomarker test for gastric cancer. Biochem Biophys Res Commun 493(3):1322–1328

Xia JT, Chen LZ, Jian WH, Wang KB, Yang YZ, He WL, He YL, Chen D, Li W (2014) MicroRNA-362 induces cell proliferation and apoptosis resistance in gastric cancer by activation of NF-kappaB signaling. J Transl Med 12:33

Xiao B, Guo J, Miao Y, Jiang Z, Huan R, Zhang Y, Li D, Zhong J (2009) Detection of miR-106a in gastric carcinoma and its clinical significance. Clin Chim Acta 400(1–2):97–102

Xin SY, Feng XS, Zhou LQ, Sun JJ, Gao XL, Yao GL (2014) Reduced expression of circulating microRNA-218 in gastric cancer and correlation with tumor invasion and prognosis. World J Gastroenterol 20(22):6906–6911

Yang Q, Jie Z, Cao H, Greenlee AR, Yang C, Zou F, Jiang Y (2011) Low-level expression of let-7a in gastric cancer and its involvement in tumorigenesis by targeting RAB40C. Carcinogenesis 32 (5):713–722

Yang Q, Zhang C, Huang B, Li H, Zhang R, Huang Y, Wang J (2013) Downregulation of microRNA-206 is a potent prognostic marker for patients with gastric cancer. Eur J Gastroenterol Hepatol 25(8):953–957

Yang M, Shan X, Zhou X, Qiu T, Zhu W, Ding Y, Shu Y, Liu P (2014) miR-1271 regulates cisplatin resistance of human gastric cancer cell lines by targeting IGF1R, IRS1, mTOR, and BCL2. Anti Cancer Agents Med Chem 14(6):884–891

Yoon K, Lee S, Han TS, Moon SY, Yun SM, Kong SH, Jho S, Choe J, Yu J, Lee HJ, Park JH, Kim HM, Lee SY, Park J, Kim WH, Bhak J, Yang HK, Kim SJ (2013) Comprehensive genome- and transcriptome-wide analyses of mutations associated with microsatellite instability in Korean gastric cancers. Genome Res 23(7):1109–1117

Yoshida S, Yamashita S, Niwa T, Mori A, Ito S, Ichinose M, Ushijima T (2017) Epigenetic inactivation of FAT4 contributes to gastric field cancerization. Gastric Cancer 20(1):136–145

Zang ZJ, Cutcutache I, Poon SL, Zhang SL, McPherson JR, Tao J, Rajasegaran V, Heng HL, Deng N, Gan A, Lim KH, Ong CK, Huang D, Chin SY, Tan IB, Ng CC, Yu W, Wu Y, Lee M, Wu J, Poh D, Wan WK, Rha SY, So J, Salto-Tellez M, Yeoh KG, Wong WK, Zhu YJ, Futreal PA, Pang B, Ruan Y, Hillmer AM, Bertrand D, Nagarajan N, Rozen S, The BT, Tan P (2012) Exome sequencing of gastric adenocarcinoma identifies recurrent somatic mutations in cell adhesion and chromatin remodeling genes. Nat Genet 44(5):570–574

Zhang Y, Guo J, Li D, Xiao B, Miao Y, Jiang Z, Zhuo H (2010) Down-regulation of miR-31 expression in gastric cancer tissues and its clinical significance. Med Oncol 27(3):685–689

Zhang X, Cui L, Ye G, Zheng T, Song H, Xia T, Yu X, Xiao B, Le Y, Guo J (2012) Gastric juice microRNA-421 is a new biomarker for screening gastric cancer. Tumour Biol 33(6):2349–2355

Zhang XL, Shi HJ, Wang JP, Tang HS, Wu YB, Fang ZY, Cui SZ, Wang LT (2014) MicroRNA-218 is upregulated in gastric cancer after cytoreductive surgery and hyperthermic intraperitoneal chemotherapy and increases chemosensitivity to cisplatin. World J Gastroenterol 20 (32):11347–11355

Zhang KC, Xi HQ, Cui JX, Shen WS, Li JY, Wei B, Chen L (2015) Hemolysis-free plasma miR-214 as novel biomarker of gastric cancer and is correlated with distant metastasis. Am J Cancer Res 5(2):821–829

Zheng B, Liang L, Wang C, Huang S, Cao X, Zha R, Liu L, Jia D, Tian Q, Wu J, Ye Y, Wang Q, Long Z, Zhou Y, Du C, He X, Shi Y (2011) MicroRNA-148a suppresses tumor cell invasion and metastasis by downregulating ROCK1 in gastric cancer. Clin Cancer Res 17 (24):7574–7583

Zhou X, Zhu W, Li H, Wen W, Cheng W, Wang F, Wu Y, Qi L, Fan Y, Chen Y, Ding Y, Xu J, Qian J, Huang Z, Wang T, Zhu D, Shu Y, Liu P (2015) Diagnostic value of a plasma microRNA signature in gastric cancer: a microRNA expression analysis. Sci Rep 5:11251

Zhu C, Ren C, Han J, Ding Y, Du J, Dai N, Dai J, Ma H, Hu Z, Shen H, Xu Y, Jin G (2014) A five-microRNA panel in plasma was identified as potential biomarker for early detection of gastric cancer. Br J Cancer 110(9):2291–2299

Chapter 7
Genetic Susceptibility Markers of Gastrointestinal Cancer

M. Kiran Kumar and Pola Sudhakar

Abstract Gastrointestinal cancer is the most common type of malignant disease with high mortality and the second foremost cause of carcinoma deaths worldwide. The frequency of gastrointestinal cancer strongly depends on ethnical and geographical characteristics. For example, the prevalence of gastrointestinal cancer is significantly high in Japan and Korea, whereas in North America and Europe the occurrence of gastrointestinal cancer is very low. Generally, gastrointestinal cancers are diagnosed in late stages due to the heterogeneous nature. By considering the heterogenicity of gastrointestinal cancer, inhibition is depending on the precise diagnosis of risk factors, the underlying cause of the disease, and the management of risk factors. Therefore, this chapter aimed to review the genetic susceptible marker of gastrointestinal cancer. Molecular studies revealed that the development of GC is from the combined effect of various factors like environment, genetic and epigenetic modifications which play a crucial role in tumorigenesis and cellular immortalization. The molecular epidemiological studies revealed that some regular genetic traits act as a genetic susceptible marker to develop GC known as single nucleotide polymorphisms (SNPs). Single nucleotide polymorphisms (SNPs) are naturally occurring genetic modifications that have a different frequency in the diversified ethnic population. There are many associate studies to analyze the genetic susceptibility of GC. Genome-wide association studies are used for the identification of various single nucleotide polymorphism in genetic susceptible markers which are responsible for gastrointestinal cancer. Gene polymorphisms become an attractive biomarker of GC due to their environment-dependent alterations. Genetic susceptibility is crucial in molecular events related to the development of gastrointestinal cancers includes mucosal shielding, immune reaction to *H. pylori* infection, carcinogen detoxification, antioxidant protection, repair of DNA injury, and capability of cell propagation. The use of SNPs as prognostic markers for individual gastrointestinal cancers is very advantageous because of the availability and quality of tumor material. In this chapter, we tried to discuss some of the important genetic

---

M. Kiran Kumar · P. Sudhakar (✉)
Department of Biotechnology, College of Science and Technology, Andhra University, Visakhapatnam, Andhra Pradesh, India

P. Veera Bramhachari, N. R. R. Neelapu (eds.), *Recent Advancements in Biomarkers and Early Detection of Gastrointestinal Cancers*, Diagnostics and Therapeutic Advances in GI Malignancies, https://doi.org/10.1007/978-981-15-4431-6_7

polymorphisms which affect gastrointestinal cancer susceptibility and how they influence malignant phenotype. The determination of genes responsible for GC susceptibility will give information for the advancement of novel GC therapeutics by studying the molecular events involved in GC carcinogenesis. This chapter includes single nucleotide polymorphisms in genes such as Cdh1, DNMT3A, PTPRCAP, PSCA, VEGF-A, XRCC1, IL-1, *HER*-2, MUC1, and MUC1.

Keywords Adenocarcinoma · Gastrointestinal cancer (GC) · Genetic susceptibility · Single nucleotide polymorphisms (SNPs) · Tumorigenesis

7.1 Introduction to Gastrointestinal Cancer

Gastrointestinal cancer is the utmost common kind of cancer with high mortality and the second foremost cause of cancer deaths worldwide (Brenner et al. 2009). Gastrointestinal carcinoma is mainly arising from the inner layer of the stomach and spread to various body parts like the liver, lungs, bones, and lymph nodes (Ruddon 2007). In Asia, gastrointestinal cancer is the most prevalent cancer, and it causes the third leading cancer deaths (Ferlay et al. 2013). Gastrointestinal carcinoma is a multifaceted disorder that is caused by the collective effect of environmental factors, host affiliated factors, genetic and biological heterogeneity. But since, several decades the incidence and mortality rates are decreased appreciably by the development of medical advancement (Ferlay et al. 2015).

7.1.1 Prevalence of GC

Cancer is considered a major cause for mortality and it increases the burden on the world due to increasing carcinogenic factors (Asombang and Kelly 2012). According to the 2012 statistics, 55% of the cancers throughout the world related to lung, breast, colon, prostate, gastrointestinal, and hepatic cancers (Zali et al. 2011). Among all the cancers gastrointestinal cancer is the fifth most common cancer around the world with 9,52,000 diagnosed cases and 7,23,000 deaths in 2012 (World Cancer Report 2014). Gastrointestinal cancer occupies the third leading cause of cancer deaths after the lung and hepatic cancer which occupies the first and second positions of cancer deaths (Lozano et al. 2012). According to the IARC 1997, large variations are observed in the incidence of gastrointestinal cancer among populations. For example, in Japan, the incidence of gastrointestinal cancer is 80 per 1,00,000 males, whereas in African states, the overall incidence of gastrointestinal cancer is only 5 per 1,00,000 people (Parkin et al. 1997). Males are more susceptible to developing stomach cancer in their lifetime which is about 1 in 95. Whereas women have a chance to develop gastrointestinal cancer which is about 1 in

154 but the risk of developing cancers for each person is affected by the other factors (American Cancer Society Cancer Facts and Figures 2019).

Gastrointestinal cancer is a predominant disease and mainly affects older people with a mean age of 60 or above. Every year out of ten about six people diagnosed with gastrointestinal cancer and it is more common in men (Parkin et al. 2005; Curado et al. 2007). The people under the age of 50 have only a 6–7% chance to develop gastrointestinal cancer, whereas less than 2% chance to develop gastrointestinal cancer for the age group below 40 (Yoshio et al. 2008). According to the American cancer society's estimations for 2019 in the United States approximately 27,510 gastrointestinal cancers are diagnosed. Among them 17,230 are men and 10,280 are women. People under 40 years of age have a chance to get stomach cancer is less than 5%. In that 5% of people under the age group, 30–39 occupies 81.1% and people under 20–29 age group occupies 18.9%. About 11,140 people have died. Among those 6800 are men and 4340 are women.

7.1.2 Global Statistics of Gastrointestinal Cancer

Worldwide the gastrointestinal cancer is the fourth most commonly occupying disease. In the western world, the incidence of gastrointestinal cancer is rapidly declining but still, GC is the second major reason of cancer-related deaths and yearly 7,40,000 deaths are recorded with the 20% 5-year survival rate. The occurrence of gastrointestinal cancer strongly depends on ethnical and geographical characteristics. For example, the prevalence of gastrointestinal cancer is significantly high in Japan and Korea, whereas in North America and Europe the occurrence of gastrointestinal cancer is very low (Crew and Neugut 2006). Japan has a more incidence of gastrointestinal cancer with an incidence of 62.7/1,00,000 than Bangladesh and India which have lower gastrointestinal cancer incidence with the incidence rates of 1.6/1,00,000 and 5.7/1,00,000, correspondingly (Fock and Ang 2010).

The occurrence of gastrointestinal cancer is different from the geographical societies. The highest prevalence of gastrointestinal cancer is found in Eastern Asia, South America, Central America, and Eastern Europe, whereas the lowermost occurrence found in Africa, North America, and Australia (Nagini 2012; Jemal et al. 2011). Hence, in high incident areas Japan, China, Korea, Eastern Europe, Central and South America, these areas are categorized into high-risk areas (Parkin et al. 2005). In North America, the major subtypes of gastrointestinal cancers are pure intestinal, pure diffuse, and mixed diffuse intestinal and their percentages are 50%, 35%, and 15%, respectively (Pisani et al. 2002). The occurrence of gastrointestinal cancer is gradually reduced in developed countries due to new inventions in medicine, but it is remaining as a serious health problem to the countries which are underdeveloped (Jahanarah et al. 2016). In the United States, the percentage of gastrointestinal cancer incidence is reduced by 1.5% per each year over the past 10 years.

In India, the occurrence rate of gastrointestinal cancer is very low when compared to the western countries and each year approximately 34,000 people diagnosed with GC with a male predominance. In India, by the end of 2020, the number of new cases will be reached to 50,000 approximately. According to the Recent Nationally Representative Survey of cancer mortality in India, the gastrointestinal carcinoma is the second common cause of cancer death among men and women (Dikshit et al. 2012). In India, the prevalence of gastrointestinal cancer is relatively high in southern areas such as Andhra Pradesh, Tamil Nadu, and Karnataka. However, recent findings demonstrate that the incidence rates of gastrointestinal cancer are increased in the North-Eastern regions of India [NCRP-2009]. Gastrointestinal cancer is the most common cancer in men than women in the Aizawl district of Mizoram (NCRP-2013). The latest reports of the National Cancer Registry Programme-2013 indicate that the incidence rates of gastrointestinal cancer for men in Aizawl and Nagaland are 64.2 and 26.2, respectively. For women, the incidence rates of gastrointestinal cancer are 31.2 in Aizawl and 12.5 in Nagaland. In Mumbai, the rates are as low as 4.2 per 1,00,000 people.

Generally, gastrointestinal cancers are diagnosed in late stages (Lee and Derakhshan 2013) due to the heterogeneous nature (Zhifang et al. 2016). The prevention and reduction of mortality rates of gastrointestinal cancers need careful attention, early detection, and proper medications (Zhifang et al. 2016). Due to the heterogenicity of gastrointestinal cancer, the prevention depends on the precise detection of risk factors, the fundamental reason of the disorder, and the management of risk factors (Yoon and Kim 2015). According to the various studies, identification and analysis of risk factors gives an effective approach for the prevention and decreasing the occurrence of GC worldwide. However, complete knowledge about the GC risk factors is essential for controlling this cancer by the plan, monitor, and evaluating national and regional states. Hence, the present study is aimed to review the genetic susceptible markers of gastrointestinal cancer.

7.2 Types of Gastrointestinal Cancer

Gastrointestinal carcinoma is a complex disorder that is categorized by the variety of different histopathological classification systems, and it is mainly classified into three pathological variants named diffuse-type, intestinal-type, and the remaining consists of mixed and indeterminate type. The diffuse-type is characterized by the development of linitis plastica which contains noncohesive single cells without gland formation and most of the times signet ring cells are existing. Hence, it is also known as signet ring cell carcinoma (Bosman et al. 2010). Diffuse GC is associated with an unfavorable prognosis due to diagnosis is often delayed until the disease advanced. *H. pylori* infection is majorly associated with intestinal-type gastrointestinal cancer. de novo diffuse-type GC is developed from the normal epithelial cells due to genetic mutations in gastrointestinal stem cells. Furthermore, in some cases, the DGC is represented by the dedifferentiated stages of IGC and also *H. pylori* contribution is present (Pilpilidis et al. 2011).

Intestinal GC is characterized by the various degrees of differentiation in the tubular or glandular components. The major intestinal GC develops from the gastrointestinal epithelium by the inflammatory changes caused due to *Helicobacter pylori* infection, and it develops chronic gastritis which leads to atrophic gastritis and finally intestinal metaplasia and dysplasia. The sequential events in intestinal GC are in the following manner, *Helicobacter* infection-, chronic inflammation-, intestinal metaplasia-, dysplasia-, adenocarcinoma. Hence, WHO recognized *Helicobacter pylori* is a class 1 carcinogen for the pathogenesis of IGC and the eradication of H.p infection is essential for the prevention of IGC. However, in both DGC and IGC, DGC has a greater chance to develop earlier in life than IGC (Crew and Neugut 2006).

7.2.1 Adenocarcinoma

Adenocarcinomas are the most frequent malignancies (90%) of the stomach which arise from the inner layer of gastrointestinal epithelium. The development of malignancies is rare from tissues like connective tissue and lymphatic tissues. Different body parts have different frequency to develop adenocarcinomas. For example, gastrointestinal cardia has a chance to develop the highest percentage of adenocarcinomas (31%), whereas the antrum and body of the stomach have only 26% and 14% chance to develop gastrointestinal cancer, respectively. Based on the histology and location, adenocarcinomas is classified and histological tumors exhibit heterogeneous appearance. Hence, the classification is mainly based on the prominent structures of tumors. Based on the gland formation and mucus secretion ability, malignancies are divided into two types. They are well-differentiated and poorly differentiated types. The majority of tubular cancers are well-differentiated and signet carcinomas are poorly differentiated. The other less common types of carcinomas are mucinous, papillary, and undifferentiated carcinomas.

7.2.2 Early Gastrointestinal Cancer

In early gastrointestinal cancer, the tumor cells restrict the stomach superficial mucosal layer and the tumors have less than 2 cm diameter, which appears as subtle lesions. The diagnosis of early gastrointestinal cancer is very crucial because the potential treatment of EGC requires endoscopic therapy is followed by an excellent diagnosis.

7.2.3 Hereditary (Familial) Gastrointestinal Cancer

Familial gastrointestinal cancer describes the chance to develop diffuse-type gastrointestinal cancer in the family members below the age of 40. The international gastrointestinal cancer linkage consortium (IGCLC) gives the measures for diagnosis. According to the IGCLC, two or more cases of diffuse-type gastrointestinal cancer in first- or second-level generation with minimum one member diagnosed before 50 years of age or pathologically identified three or more cases in the first- and second-level family members irrespective of the age. Among the family, one-third members have a germline mutation of the CDH1 gene. The affected family members also have a greater risk to develop breast and colon cancer.

7.2.4 Lymphoma

Gastrointestinal lymphomas are two types; they are B or T cell types. The B cell gastrointestinal lymphomas are developed primarily from the stomach particularly in the mucosal-associated lymphoid tissue (MALT), and these are considered as low-grade tumors. These lymphomas are highly favorable for clinical therapies, but they have a high frequency of transformation.

7.3 Causes of Gastrointestinal Carcinoma

GC is considered as a complex disease because both the environmental and genetic factors have a major part in the growth of GC. Gastrointestinal carcinoma is highly prevalent in the lower socioeconomic classes and is frequently detected in the advanced conditions (Carcas 2014). Diverse environmental factors that enhance gastrointestinal cancer risk include *Helicobacter pylori* and EBV infection, more salt and more nitrogen foods, tobacco, pre-malignant stomach lesions and genetic factors. All the described factors are referred to as gastrointestinal cancer risk factors (González and Agudo 2012). Among all the above-mentioned cases, the *Helicobacter pylori* infection is the major cause for developing gastrointestinal cancer which accounts for approximately 60% of cases (Fiona and Martin 2011). Other common causes for gastrointestinal cancers are packed vegetables, smoking, and genetic mutations (World Cancer Report 2014). Molecular studies revealed that the development of GC is from the combined effect of various factors like environment, genomic, and epigenetic modifications which show a vital role in tumorigenesis and cellular immortalization (World Cancer Report 2014).

Single nucleotide polymorphisms (SNPs) are naturally occurring genetic modifications that have a different frequency in the diversified ethnic population. Researchers focus on the identification of novel genetic susceptibility markers for all the types of gastrointestinal carcinomas. SNPs can modify the expression pattern of genes and can alter the function of a gene, which leads to an increased risk of diverse diseases, including cancer. There are numerous examples for the existence of polymorphic genes which increase susceptibility to GC (Pinheiro et al. 2010). Nowadays, there is a possibility of identification and getting information of unexplored SNPs within a large number of genes by using advanced technologies like Genome-Wide Association Studies (GWAS) and high-throughput genomic investigation. These new methods for the detection of SNPs simultaneously give perceptions for the pathogenesis of GC. The development of the malignant disease is mainly by the cumulative effect, particularly by the genetic polymorphism, ethnicity, and exposure to environmental risk factors (Saeki et al. 2011).

In recent years, genetic markers show a significant role in the identification and management of patients with gastrointestinal carcinomas especially colorectal cancer, gastrointestinal stromal tumors, gastrointestinal and gastro-esophageal junction cancers. In 2003 and 2007, the European Group of Tumour Markers (EGTM) establish guidelines for the use of biomarkers in CRC (McLean and El-Omar 2014). This chapter provides new inventions on the use of biomarkers in gastrointestinal and gastro-esophageal junction cancers and gastrointestinal stromal tumors.

7.4 Genetic Susceptible Markers of Gastrointestinal Cancers

The development of gastrointestinal cancer is associated with multiple factors such as gastritis, gastrectomy (Duffy et al. 2007), *Helicobacter pylori* infection (Rugge et al. 2014), and genetic susceptibility factors (Gatti et al. 2004). Genetic factors play an important role in the development of gastrointestinal cancer. The familial clustering phenomenon of gastrointestinal cancer reveals that only a small fraction of people is affected after they exposed to the same environment. This phenomenon indicated that environmental exposure plays a major role in genetic susceptibility which leads to gastrointestinal cancer development in individuals (Xie et al. 2014). The epidemiological studies also reveal that only a small percentage of people who exposed to an environment with high incidence rates of gastrointestinal cancer are affected. These studies suggested that the chance of an individual to get gastrointestinal cancer depends on the individual's genetic susceptibility. In this chapter, we aimed to summarize the relationship between genetic polymorphisms and gastrointestinal cancer susceptibility.

Germline alterations in sequence of DNA are supposed to represent the main feature of a tendency to most complex traits, such as cancer (Milne et al. 2009). Genetic diseases triggered through the gradual accumulation of modifications in genes that regulate the differentiation, growth, and DNA repair can lead to the development of gastrointestinal cancer (Kelly et al. 2009). A small percentage of people only develop GI cancers based on their hereditary component and it is proved through the well-studied genetic disorders and the family history associated risk factors (Grady and Markowitz 2002). Approximately 5% of hereditary genetic disorders are due to strong mutations evident with well-studied experimental demonstrations (Garber and Offit 2005). 20–25% of genetic disorders are associated with a hereditary component, which is not established until now (Jasperson et al. 2010). Several gastrointestinal cancers develop due to mutations in one gene, and this type of cancers are less carrying but develop persistently than the other cancers which are developed by combination with well-studied genetic disorders (Kelly et al. 2009). The single gene polymorphisms (SNPs) in genes which are participated in the regulation of metabolic pathways or the genes can be controlled by environmental influences (Jasperson et al. 2010). Mutations in multiple susceptible loci can also lead to the development of cancers by inducing additive effects (Grady and Markowitz 2002). This chapter discourses the genomics of the well-studied hereditary cancers of the GI tract.

The molecular epidemiological studies revealed that some common genetic traits act as a genetic susceptible marker to develop GC known as single nucleotide polymorphisms (SNPs) (Oliveira et al. 2006). Gastrointestinal carcinogenesis is also depending on the host genetic risk factors. Hence, gene polymorphisms become an attractive biomarker of GC due to their environment-dependent alterations. Genetic susceptibility crucial in molecular events related to the development of gastrointestinal cancers includes mucosal shielding, immune reaction to *H. pylori* infection, carcinogen detoxification, antioxidant protection, repair of DNA injury, and capability of cell propagation (Yin et al. 2009). The use of SNPs as prognostic markers for gastrointestinal cancers is very advantageous because of the accessibility and quality of tumor material, and they can be determined independently and easily evaluated from individual blood samples. There are many associate studies to analyze the genetic susceptibility of GC. For example, genome-wide association studies are used for the identification of various single nucleotide polymorphism in genetic susceptible markers which are responsible for gastrointestinal cancer. HDGC is a sporadic autosomal dominant disease, and it is produced by the germline mutations in the CDH1 gene, which translates cell adhesion molecule known as E-cadherin. 70–80% of gastrointestinal cancers are developed by the mutations in the CDH1 gene.

GC also develops other types of familial cancers, including Lynch syndrome, familial adenomatous polyposis, Peutz–Jeghers syndrome, and Li–Fraumeni syndrome. The prevalence of gastrointestinal cancer is 2.9 times higher in individuals who are having germline mutations in the MLH1 gene. A study by Hansford et al. (Gonzalez et al. 2002) reported that 12% of CDH1-negative HDGC families have germline mutations in tendency genes including CTNNA1, BRCA2, STK11, PALB2, ATM, MSR1, and SDHB. All these genes can sense the development of gastrointestinal cancer in families and provide molecular evidence of tumorigenesis (Hansford et al. 2015). DNA methyltransferase 3A is responsible for the genomic methylations and also essential for the differentiation of stem cells during development in mammals (Ding et al. 2008). Fan et al. (Yurgelun and Boland 2017) reported that the polymorphism of gene DNMT3A-448 A>G is involved in the development of gastrointestinal cancer by acting as a genetic susceptible marker for GC. Ding et al. (Fan et al. 2010) reported that gastrointestinal carcinogenesis involves the de novo expression of the DNMT3A gene.

Protein Tyrosine Phosphatase Receptor Type C-Associated Protein (PTPRCAP) is participated in the stimulation of Src family kinases (SFKs) (Motoya et al. 1999) and the activated SFks play a key role in the interruption of the epithelial adherin junctions by dislocating the E-cadherins in membranes (Avizienyte et al. 2002). PSCA gene also reported as a genetic susceptible marker for gastrointestinal cancer (Sakamoto et al. 2008a). PSCA is overexpressed in differentiating gastrointestinal epithelial cells to inhibit the cell proliferation and its silenced form mostly found in gastrointestinal carcinomas. Lu et al. (Lu et al. 2010) reported that two polymorphisms (rs 2976392 and rs 2294008) in the PSCA gene lead to gastrointestinal carcinogenesis. The VEGF (Vascular Endothelial Growth Factor) gene has been identified in many genome-wide association studies as a genetic susceptible marker. VEGF is important for the progression of various tumors including GC by acting as a key factor in angiogenesis (Ke et al. 2008). Several studies reported that the VEGF 634 G>C polymorphism involved in the increasing risk to form GC (Guan et al. 2009).

The polymorphism 1612 G>A in the 3′-UTR of VEGF is associated with the deregulation of affected genes and thereby increasing the risk of gastrointestinal cancer (Tahara et al. 2009). The gene XRCC1 (X-ray Repair Cross-Complementing Group 1) is involved in the maintenance of integrity and DNA nucleotide composition, and it is important for the normal functioning of the cell. XRCC1 is participated in the base excision repair mechanism that repairs the single nucleotide changes produced by the ionizing radiations, alkylating agents, and metabolic toxins (Caldecott et al. 1995). The XRCC1-77 T>C polymorphism in promoter region is correlate with human cancer known as non-small cell lung cancer (Hao et al. 2006). Corso et al. (2009) reported that the relation between the XRCC1 77 T>C polymorphism and the increased risk of gastrointestinal carcinoma. Hence, the polymorphism of XRCC1 can be used as a host genetic susceptible factor for gastric carcinoma.

Host genetic features act as a key element in the increased risk for the development of cancer, and the associations of various polymorphisms on diversified genes and their products interact with environmental factors and provide important information to explain the multiple risks in diversified populations. EI-Omar et al. (2000a) reported that the interaction of precise gene variants increases the risk of gastric carcinoma. The meta-analysis of individual cytokine gene polymorphism in GC susceptibility reveals that the association between specific variants of IL1RN VNTR, IL1B-511, and IL10-1082 gene polymorphisms increase the GC risk. The interleukin-1 beta IL1B-31T (rs 1143627) and IL-1 receptor antagonist IL1RN2/2 genes are linked with an enhanced risk of both chronic hypochlorhydria and GC by altering IL-1 concentration in the stomach. Genetic polymorphism along with the susceptibility of cancer also affects the tumor phenotype. Total genome expression studies are useful for the identification of new genes involved in invasion, metastasis, and potential prognostic factors. Sequential analysis of gene expression studies identifies the several genetic susceptibility genes such as CDH1, APOE, FUS, COL1A1. COL1A2, GW112, and MIA.

7.4.1 CDH1 Gene

The CDH1 gene is present on the 16q 22.1 chromosome of human and it contains 16 exons, which transcribed into 4.5 kb m-RNA and translates into E-cadherin (Bussemakers et al. 1994). E-cadherin is a calcium-dependent cell adhesion molecule which plays an important role in maintaining polarity and differentiation of cells by forming adherin junctions and desmosomes (Stemmler 2008). E-cadherin is a glycoprotein which contains three domains known as small cytoplasmic domain, transmembrane domain, and large extracellular domain. Five tandemly repeated domains are present in the extracellular domain and they are named as EC1- EC5 (Takeichi 1995). The extracellular domains of cadherins involved in cell–cell interactions by forming homophilic dimerization.

The cytoplasmic domain of E-cadherin contains three different types of catenins known as α, β, and γ. These catenins are involved in the anchoring of cadherins by establishing the interaction between the cytoplasmic domain of cadherin and the actin in cytoskeleton (Gumbiner and Mccrea 1993). E-cadherin is predominantly expressed in the epithelial cells and makes strong adherin junctions thereby; suppress the invasion (Yagi and Takeichi 2000). Germline mutations of CDH1 allele produces diffuse-type gastric carcinoma by the inhibition of E-cadherin second allele is by the methylations, mutations. Furthermore, researchers reported that the cancer cells migrate to various body parts and make changes among the cancer cells and the constituents of extracellular matrix (Valastyan and Weinberg 2011). This leads to tumor progression by altering the cell–cell adhesions and cell–matrix adhesions. E-cadherin and the cadherin–catenin complex in the cytoplasmic side of the epithelial cells involved in the various signalling pathways include Wnt signalling, Rho GTPases, and NF-κB. Hence, mutations in E-cadherin affect these signalling pathways by influencing the cell polarity, cell survival, invasion, and migration in gastric carcinogenesis.

E-cadherin also exhibits various partners for making interaction in the cytoplasmic adhesion complex with the actin filament. The Epithelial Mesenchymal Transition (EMT) process is by the several signalling pathways such as Wnt signalling, Rho GTPases, and EGFR (Cavallaro and Christofori 2004). The inhibition of E-cadherin expression on epithelial cells leads to decreasing the polarity of a cell and enhances the migratory and invasive development characteristics by the initiation of active signals for EMT (Garcia de Herreros and Baulida 2012). The WNT gene family proteins are involved in a signalling pathway for embryonic development and oncogenesis. This signalling can be subdivided into two types: β-catenin-dependent signalling or canonical Wnt signalling. Another type of signalling is β-catenin-independent signalling, known as non-canonical Wnt signalling. The WNT gene family involves glycogen synthase kinase-3beta (GSK-3beta) molecules, beta-

and gamma-catenins, and APC. Beta-catenin binds directly to the intracellular domain of E-cadherin and alpha-catenin by interacting with cytoskeletal actin through the APC protein. Beta-catenin is also a transcriptional coregulator and is persistently targeted for proteasomal degradation by the APC/Axin/GSK3b complex when the pathway is inactive. In the canonical pathway, the Wnt protein subtype binds to receptors on the cell membrane and inactivates the APC/Axin/GSK3b; thereby degradation of b-catenin is prevented which leads to increased levels of free cytoplasmic beta-catenin. The free b-catenin migrates into the nucleus where it forms a complex with LEF-1/TCF which is capable of promoting transcription of other genes involved in proliferation (Staal et al. 2008).

Maximum level concentration of β-catenins in the cytoplasm immediately translocate into the nucleus and binds to the TCF/LEF1 elements. Furthermore, stimulates the Wnt target genes expression, including CD44, c-MYC, cyclin D1, and MMP7 (Moon et al. 2004). Activation of these genes enhances the rate of cell proliferation and induces tumor formation. E-cadherin expression on cell membrane inhibits the Wnt β-catenin signalling pathway by sequestering the β-catenin at the sites of cytoplasmic domain and cytoskeleton junction. Hence, the cytoplasmic domain is important for the inhibition of Wnt β-catenin-mediated expression of gene (Gottardi et al. 2001). Various cellular systems demonstrated that the sequestration of β-catenin by E-cadherin can compete with the β-catenin/TCF-mediated transcriptional activity of the canonical Wnt signalling pathway (Cavallaro and Christofori 2004).

Besides the Wnt signalling, there is another pathway induced by the E-cadherin extracellular domain (Suriano et al. 2003) which is mediated by the enhanced RhoA activity and leads to acquire high migration ability. EGFR (Epidermal Growth Factor) plays a major role in the stimulation of RhoA by an E-cadherin-mediated pathway (Bremm et al. 2008). Hence, mutations in the extracellular domain of E-cadherin lead to wrong interaction with EGFR and activate the EGFR, which leads to increase the motility of cell by RhoA activation (Mateus et al. 2007). However, loss of E-cadherin also releases P120-catenin which activates the Rac1-MAPK (Mitogen-Activated Protein Kinase) signalling pathway and induces the overexpression of RhoA, Rac1, and Cdc42 (Pan et al. 2004). The above signalling molecules are thought to play a critical role in the organization of cytoskeleton, motility of cell, and promotion of cell growth (Heasman and Ridley 2008).

H. pylori infection induces the gastric cancer by inflammation-associated carcinogenesis during inflammation of epithelial cells is regulated by the NF-κB (Karin and Greten 2005). In mammals, activation of canonical NF-κB signalling pathway is mainly by the dimerization of P65:P50. Under normal conditions the NF-κB is inactivated by the IFK. Upon inflammation, the mediators such as cytokines and microbial or endogenous molecules induce the release of P65:P50 by phosphorylation of IKB by the IKK complex. The free P^{65}:P^{50} heterodimer moves into the nucleus and triggers the expression of response-specific genes includes Bcl-2, IL-6, and TNF (Ben-Neriah and Karin 2011). The expression of these genes increases the proliferative ability and decreases the apoptosis ability, thereby enhancing the chance to develop inflammation-associated tumour growth. In a cell, it is evident that the hyperexpression of E-cadherin can decrease the NF-κB stimulation, whereas loss of E-cadherin induces the activity of NF-κB (Kuphal et al. 2004). NF-κB suppression is mediated by the catenin and E-cadherin complex (Solanas et al. 2008). Hence, the stimulation of NF-κB through the down regulation of E-cadherin gives information for the *H. pylori* infection-related gastrointestinal cancer development.

7.4.2 DNMT3A Gene

Eukaryotes have three types of DNA methyltransferases called as DNMT1, DNMT2, DNMT3A, and DNMT3B. DNMT1 is involved in the maintenance of the pre-existing methylation patterns on DNA during replication; hence, it is called maintenance methyltransferase (De Carvalho et al. 2012). Whereas the DNMT3A and DNMT3B are involved in the formation of new methylation patterns during embryogenesis; hence, they are called de novo methylases (Liu et al. 1998). Among all other DNA methylases DNMT3A is crucial for the development of several cancers including gastrointestinal cancer (Li 2002). DNA methylation plays an important role in epigenetic inheritance of DNA. If there are any abnormalities in DNA methylation, that will lead to the development of cancer (Robertson et al. 1999). Abnormal DNA methylation in gastric epithelial cells leads to alter the expression of tumor suppressor genes which are involved in the carcinogenesis (Park et al. 2006).

The cell proliferation and differentiation in gastrointestinal epithelium are controlled by the intracellular factors called cell cycle regulators (Kang et al. 2008). The cell division is negatively controlled by the inhibition of CDK4 (INK4)-CDK4/6 Cyclin D-Rb-E2F pathway (Neureiter et al. 2006). The inactivation of INK4 enhances the formation of active CDK4/6-cyclin D complex, which is involved in cell proliferation. The INK4 family includes P16INK4A, P15 INK4B, and P18 INK4C (Canepa et al. 2007). The risk of developing cancer in gastric intestinal epithelial cells significantly increases with the P16 deregulation (Sherr and Roberts 1995) and Rho A-mediated inactivation of INK4 family members. This leads to the loss of cell cycle regulation particularly at G1-S transition, and it indicates that the INK4 family proteins have a significant role in gastrointestinal epithelial cell proliferation (Sun et al. 2004). In addition to the silencing of INK4 members, RhoA also involves in the development of cancers by inducing promoter hypermethylation (Zhang et al. 2009). DNMT3A induces the gastrointestinal carcinoma by methylating the P18 INK4C gene product and decreases the expression of P18 INK4C; it leads to the dysregulation of G1-S check point. The loss of G1-S regulation leads to unregulated cell proliferation and induces gastrointestinal carcinoma. All these findings are useful for the development of new drugs and therapies to treat gastrointestinal cancer by specifically target DNMT3A.

7.4.3 PTPRCAP Gene

The Protein Tyrosine Phosphatase Receptor Type C Associated Protein (PTPRCAP) also called as CD45-AP is involved in the carcinogenesis by acting as a positive regulator for the protein tyrosine phosphatase. The PTPRC carries signals intracellularly by activating the Src family kinases such as SFK. The protein tyrosine phosphatase activity of PTPRC can dephosphorylate the inhibitory phosphate groups leading to activation of SFK (Takeda et al. 2004). The phosphatase activity of PTPRC is activated by the interaction with PTPRCAP transmembrane domain. The inactivation of SFK is by the phosphorylation of inhibitory tyrosine residues at carboxy terminal (Barraclough et al. 2007). Both the protein tyrosine kinases and phosphatases interact with each other and regulate the signal transduction cascades for cell proliferation (Hyoungseok et al. 2009). The deregulated protein tyrosine phosphatase is involved in the progression of carcinoma (Hunter and Cooper 1985). The diffuse-type gastrointestinal cancer is associated with the polymorphism in the promoter of PTPRCAP gene at the position of −309 (Kirsch et al. 2009); hence, it can be used as genetic susceptible marker for the gastrointestinal cancer. Human epithelial cell carcinomas are characterized by the overproduction of SFK protein (Matsuda et al. 1998).

7.4.4 PSCA Gene

Prostate-Specific Cell Surface Antigen (PSCA) is a glycoprotein made up of 123 amino acids, and the PSCA gene is present on the 8q24.2 chromosome (Sakamoto et al. 2008b). PSCA protein used for the intracellular signal transduction due to the presence of GPI-anchored proteins. The extracellular domain of PSCA has a microdomain which contains high amounts of glycolipids, cholesterol, and lipidate proteins; hence, it exists on the extracellular lipid rafts of the cell membrane (Reiter et al. 1998). The PSCA gene is a tumor suppressor gene, and it is associated with susceptibility of gastrointestinal carcinomas by altering the properties of cell–cell adhesion and proliferation (Sharom and Radeva 2004). Hence, the PSCA gene can be used as genetic susceptible marker for gastrointestinal cancer (Summy and Gallick 2003). The polymorphism in PSCA gene such as rs2294008 makes the gene more expressive; this leads to the carcinogenesis (Hruska et al. 2009). The polymorphism changes the first amino acid of PSCA methionine into threonine; this change leads to the premature termination of the 9 amino acid length truncated PSCA protein (Fu et al. 2012). The down regulation of PSCA enhances the cell growth inhibitory properties in gastrointestinal epithelial cells (Summy and Gallick 2003).

7.4.5 VEGF-A Gene

Vascular endothelial growth factor A (VEGF-A) is the most formidable factor for the neoangiogenesis. Neoangiogenesis is the process of formation of new blood vessels from the pre-existing precursor endothelial cells, and this is the pathological symptom of inflammation, epithelial ulcers, and the carcinogenesis and metastasis (Yancopoulos et al. 2000). VEGF is a secreted protein and it also plays an important role in increasing the permeability of blood vessels (Senger et al. 1983). Carcinogenic tissue is characterized by the high presence of VEGF-A. This high content of VEGF is due to secretion by the cancer cells and also from the fibroblast and inflammatory cells which constitute the stroma of a tumor (Senger et al. 1983).

The elevated levels of VEGF expression in stomach primarily indicated the peptic lesion healing (Fukumura et al. 1998). In addition, high levels of VEGF also indicate the presence of gastrointestinal adenocarcinoma, which is accomplished by the increased intra-tumoral microvessel density (Jones et al. 2001). Furthermore, VEGF overexpression in gastric malignant lesions like chronic atrophic gastritis and intestinal metaplasia suggests that the alterations of VEGF expression may also involve in the process of gastric carcinogenesis (Maeda et al. 1999). The intestinal-type of gastric cancers is mere dependent on the angiogenesis compared to diffuse-type gastric cancer. The levels of VEGF-A and the number of blood vessels are correlated significantly in the gastrointestinal carcinomas (Feng et al. 2000).

A polymorphism in VEGF gene 3′ UTR sequences associated with the elevated VEGF levels in serum (Yamamoto et al. 1998). The 3′ UTR sequences of VEGF involved in the stabilization of m-RNA, and it is also involved in the induction of VEGF in hypoxic condition. Researchers identify the genes named as Hu family; their products bind to the 3′ UTR AU-rich sequences of several m-RNAs including VEGF m-RNA (Renner et al. 2000). The Hu protein to 3′ UTR alters the confirmation of m-RNA and makes it resistant to RNAase attack. These alterations in 3′ UTR leads to VEGF gene polymorphism which affects the respective functions of the gene (Awata et al. 2002). The polymorphism in the VEGF-A 3′ UTR sequence induces m-RNA conformational integrity and causes overexpression of VEGF; finally, it leads to gastrointestinal cancers.

7.4.6 XRCC1 Gene

The X-ray repair cross-complementing group (XRCC) is the major protein involved in the repair mechanisms of DNA. The XRCC1 acts as a scaffolding protein, which directly interacts with the DNA polymerase β, ADP-ribose polymerase (PARP), DNA ligase III and forms a complex which is involved in the base excision repair mechanism (Ramamoorti et al. 2001). Several external and internal mutagens such as ionizing radiations, alkylating agents, deaminating agents, and reactive oxygen radicals cause the DNA damage, and this type of DNA damages are repaired by the base excision repair mechanism (Caldecott et al. 1996). The XRCC1 protein independently recognizes the nicks or gaps in the damaged DNA and induces the repair mechanism by recruiting DNA polymerase β (Christmann et al. 2003). The DNA repair mechanisms might play a key role in the development of gastrointestinal cancer. Hence, it can be referred as genetic susceptible markers of gastrointestinal cancers. Two different point mutations in the XRCC1 gene conserved sites make the gene polymorphic. The two-point mutations are substitution type. One of the mutations is C to T substitution is located at 194 codon in exon 6 and other mutation is G to A substitution located at 399 in exon 10 which leads to the alteration of amino acids arginine to tryptophan and arginine to glutamine, correspondingly. These alterations induce the carcinogenesis in gastrointestinal track (Butkiewicz et al. 2000). The amino acid alterations change the repair capacity of XRCC1 and increase the chance of DNA damage which leads to carcinogenesis (Marintchev et al. 1999).

7.4.7 IL-1 Gene

Interleukin-1 (IL-1) is involved in the cell proliferation and differentiation by acting as a pro-inflammatory chemokine (Huang et al. 2005). The IL-1 family have three types of interleukins known as IL-1α, IL-1β, and IL-1 receptor antagonist (IL-1RA). IL-1α and IL-1β are synthesized from the various genes but they show functional similarities by binding with the same receptor and elicit same biological responses (England et al. 2014). IL-1 is present in cytoplasm, and it regulates the expression of several genes which are involved in the tumor induction, metastasis, and angiogenesis (Fanjul-Fernández et al. 2010; Rahim et al. 2014; Akdis et al. 2011; Liacini et al. 2002). 1α and IL-1RA have structural homology with both IL-1α and IL-1β, and it binds to the IL-1 receptor type 2 without delivering an activation signal. IL-1 acts as an antagonist to the IL-1α and IL-1β. The binding of IL-1RA induces the molecular reorganization of receptor and acts as an inhibitor for IL-1 (Wei et al. 2015). IL-1RA involves in the down regulation of IL-6 and IL-8 in pancreatic carcinomas (Apte and Voronov 2002) and VEGF in gastrointestinal carcinoma (Matsuo et al. 2004). The polymorphism in the IL-1β and IL-1RN increases the risk of gastrointestinal carcinogenesis. Two linked IL-1β polymorphism such as 511 C>T and 31 T>C induces the overexpression IL-1 h, and it enhances the risk of gastrointestinal carcinoma (Ma et al. 2009).

7.4.8 HER-2 Gene

The Human Epidermal Growth Factor Receptor 2 (HER) is a 185 KDa, transmembrane glycoprotein receptor. The HER-2 acts as proto-oncogene, and it is encoded by gene ERBB2 which is located on 17q11.2-12 chromosome (El-Omar et al. 2000b). The epidermal growth factor receptor family includes four types of proteins. They are HER-1, HER-2, HER-3, and HER-4. The HER-1 is also called as EGFR, HER-2 and HER-3 are called as ErbB-3, and HER-4 is called as ErbB-4. All the HER family members have a similar molecular and structural organization. HERs are transmembrane proteins that consist of an extracellular ligand binding domain, cytoplasmic domain with tyrosine kinase activity, and a transmembrane domain. The binding of ligands with the extracellular domains induces the signal transduction through the activation of activated mitogen-protein kinase, phosphoinositide-3 kinase, phospholipase-C, protein kinase-C, signal transducers, and transcription factors. The activation of transcription factors involves in the induction of several responses such as proliferation, apoptosis, adhesions, migration, and differentiation (Baselga et al. 1996). The abnormal HER-2 expression induces the cell proliferation and inhibits the apoptosis, which leads to carcinogenesis (Olayioye 2001). This abnormal HER-2 can induce cancers in different types of tissues including breast, kidneys, heart, and gastrointestinal tract (Baselga et al. 1996).

7.4.9 MUC1 Gene

The mucin genes are mainly involved in the protective function of gastric mucosa. There are different subtypes of mucin genes, such as MUC1, MUC2, MUC5AC, MUC6, and trefoil peptide family (Bafna et al. 2010). MUC1 gene product is a 2000 kDa transmembrane glycoprotein which interacts with the bicarbonate ions to protect gastric mucosa by forming mucus-bicarbonate barrier (Wang and El-Bahrawy 2015). During post-translational modifications, the MUC1 peptide is fragmented by self-proteolysis into N-terminal and C-terminal subunit named as MUC1-N and MUC1-C, correspondingly. These two subunits are non-covalently attached to each other and present on the external side of the epithelial cell membrane. The transmembrane MUC1-N has several sites for glycosylation, and it gives protection to the cells non-specifically (Nath and Mukherjee 2014). The MUC1-C has a transmembrane domain and a cytoplasmic domain and it is involved in intracellular signal transduction. The cytoplasmic domain of MUC1-N contains many phosphorylation sites and a single β-catenin binding site. The formation of MUC1-N cytoplasmic tail and β-catenin complex is induced by the phosphorylation of Thr residues in the TDRSPYEKV sequence of cytoplasmic tail, and it leads to the activation of cell cycle regulating gene p53 by nuclear localization of the complex (Sandra 2001). Several studies demonstrate that the expression of MUC1 is significantly increasing with prognosis of cancer. Hence, MUC1 is considered as an oncoprotein, and it can be used as a genetic susceptible marker for the gastrointestinal carcinomas.

7.5 Conclusions and Future Perspectives

Gastrointestinal carcinoma is a complex disease that is caused by the combined effect of environmental factors, host affiliated factors, genetic and biological heterogenicity. The lifestyle and dietary habits of individuals, associated with genetic susceptibility and molecular changes developed throughout the lifetime, are the basis for the carcinogenesis of GC. Abundant research has been accomplished to find molecular markers for GC. Understanding the consequences of these mutations in susceptibility is attaining importance in cancer research to develop new therapeutic and preventive measures. Furthermore, the molecular pathways are required to know the causes of GC and make a possibility to achieve the best clinical methods to assure an accurate diagnosis and effective treatment. Attaining a comprehensive molecular understanding of the several genomic abnormalities related to GC will be crucial to enhance the results of patients. Recent research has seen significant improvement in decoding the genetic information of GC by finding novel molecular mechanisms involved in cellular pathways that are associated with gastrointestinal carcinoma and development. The identification and analysis of GC susceptibility risk factors give an effective approach for the prevention and decrease the occurrence GC in the future. The prevention and the development of new therapeutics for GC are possible by a systematic unveiling of novel molecular pathways involved in GC carcinogenesis, and it is the vital program in medical research to overcome the socioeconomic burden of cancer deaths.

Acknowledgments Kiran Kumar M. and Sudhakar Pola are grateful to College of Science and Technology, Andhra University, Visakhapatnam, India, for providing necessary facilities to carry out the research work and for extending constant support.

Conflict of Interest The authors declare that there is no potential conflict of interest.

References

Akdis M, Burgler S, Crameri R et al (2011) Interleukins, from 1 to 37, and interferon-γ: receptors, functions, and roles in diseases. J Allergy Clin Immunol 127(3):701–721

American Cancer Society Cancer Facts & Figures (2019)

Apte RN, Voronov E (2002) Interleukin-1-a major pleiotropic cytokine in tumor-host interactions. J Semin Cancer Biol 12(4):277–290

Asombang AW, Kelly P (2012) Gastric cancer in Africa: what do we know about incidence and risk factors. Trans R Soc Trop Med Hyg 106:69–74

Avizienyte E, Wyke AW, Jones RJ et al (2002) SRC-induced deregulation of E-cadherin in colon cancer cells requires integrin signaling. Nat Cell Biol 4:632–638

Awata T, Inoue K, Kurihara S, Ohkubo T, Watanabe M, Inukai K, Inoue I, Katayama S (2002) A common polymorphism in the 5′- untranslated region of the *VEGF* gene is associated with diabetic retinopathy in type 2 diabetes. Diabetes 51:1635–1639

Bafna S, Kaur S, Batra SK (2010) Membrane-bound mucins: the mechanistic basis for alterations in the growth and survival of cancer cells. Oncogene 29:2893–2904

Barraclough J, Hodgkinson C, Hogg A, Dive C, Welman A (2007) Increases in c-Yes expression level and activity promote motility but not proliferation of human colorectal carcinoma cells. Neoplasia 9:745–754

Baselga J, Tripathy D, Mendelsohn J, Baughman S, Benz CC, Dantis L, Sklarin NT, Seidman AD, Hudis CA, Moore J, Rosen PP, Twaddell T, Henderson IC, Norton L (1996) Phase II study of weekly intravenous recombinant humanized anti-p185HER2 monoclonal antibody in patients with HER2/neu-overexpressing metastatic breast cancer. J Clin Oncol 14:737–744

Ben-Neriah Y, Karin M (2011) Inflammation meets cancer, with NF-κB as the matchmaker. Nat Immunol 12:715–723

Bosman FT, Carneiro F, Hruban RH, Theise N (2010) WHO classification of tumours of the digestive system, 4th edn. IARC, Lyon

Bremm A, Walch A, Fuchs M, Mages J, Duyster J, Keller G et al (2008) Enhanced activation of epidermal growth factor receptor caused by tumor-derived E-cadherin mutations. Cancer Res 68:707–714

Brenner H, Rothenbacher D, Arndt V (2009) Epidemiology of stomach cancer. Methods Mol Biol 472:467–477

Bussemakers MJG, van Bokhoven A, Völler M, Smit FP, Schalken JA (1994) The genes for the calcium-dependent cell adhesion molecules P- and E-cadherin are tandemly arranged in the human genome. Biochem Biophys Res Commun 203:291–1294

Butkiewicz D, Rusin M, Harris CC, Chorazy M (2000) Identification of four single nucleotide polymorphisms in DNA repair genes: *XPA* and *XPB (ERCC3)* in polish population. Hum Mutat 15:577–578

Caldecott KW, Tucker JD, Stanker LH, Thompson LH (1995) Characterization of the XRCC1-DNA ligase III complex in vitro and its absence from mutant hamster cells. Nucleic Acids Res 23:4836–4843

Caldecott KW, Aoufouchi S, Johnson P, Shall S (1996) XRCC1 polypeptide interacts with DNA polymerase beta and possibly poly (ADP-ribose) polymerase, and DNA ligase III is a novel molecular 'nick-sensor' *in vitro*. Nucleic Acids Res 24:4387–4394

Canepa ET et al (2007) INK4 proteins, a family of mammalian CDK inhibitors with novel biological functions. IUBMB Life 59:419–426

Carcas LP (2014) Gastric cancer review. J Carcinog 13:1–14

Cavallaro U, Christofori G (2004) Multitasking in tumor progression: signaling functions of cell adhesion molecules. Ann N Y Acad Sci 1014:58–66

Christmann M, Tomicic MT, Roos WP, Kaina B (2003) Mechanisms of human DNA repair: an update. Toxicology 193:3–34

Corso G, Marrelli D, Pedrazzani C et al (2009) Gastric cardia carcinoma is associated with the promoter −77T>C gene polymorphism of X-ray cross-complementing group 1 (XRCC1). J Gastrointest Surg 13:2233–2238

Crew KD, Neugut AI (2006) Epidemiology of gastric cancer. World J Gastroenterol 12:354–362

Curado MP, Edwards B, Shin HR, Storm H, Ferlay J, Heanue M et al (2007) Cancer incidence in five continents, vol IX. IARC Press, Lyon

De Carvalho DD, Sharma S, You JS, Su SF, Taberlay PC, Kelly TK, Yang X, Liang G, Jones PA (2012) DNA methylation screening identifies driver epigenetic events of cancer cell survival. Cancer Cell 21:655–667

Dikshit R, Gupta PC, Ramasundarahettige C, et al; the Million Death Study Collaborators (2012) Cancer mortality in India: a nationally representative survey. Lancet 379:1807–16

Ding WJ, Fang JY, Chen XY, Peng YS (2008) The expression and clinical significance of DNA methyltransferase proteins in human gastric cancer. Dig Dis Sci 53:2083–2089. https://doi.org/10.1007/s10620-007-0145-2. [PMID: 18253830]

Duffy MJ, van Dalen A, Haglund C et al (2007) Tumour markers in colorectal cancer: European Group on Tumour Markers (EGTM) guidelines for clinical use. Eur J Cancer 43:1348–1360. https://doi.org/10.1016/j.ejca.2007.03.021

El-Omar EM, Carrington M, Chow WH, McColl KE, Bream JH, Young HA, Herrera J, Lissowska J, Yuan CC, Rothman N, Lanyon G, Martin M, Fraumeni JF Jr, Rabkin CS (2000a) Interleukin-1 polymorphisms associated with increased risk of gastric cancer. Nature 404:398–402

El-Omar EM, Carrington M, Chow WH et al (2000b) Interleukin-1 polymorphisms associated with increased risk of gastric cancer. Nature 404:398–402. Erratum in: Nature 2001 Jul 5; 412:99

England H, Summersgill HR, Edye ME et al (2014) Release of interleukin-1α or interleukin-1β depends on mechanism of cell death. J Biol Chem 289(23):15942–15950

Fan H, Liu D, Qiu X, Qiao F, Wu Q, Su X, Zhang F, Song Y, Zhao Z, Xie W (2010) A functional polymorphism in the DNA methyltransferase-3A promoter modifies the susceptibility in gastric cancer but not in esophageal carcinoma. BMC Med 8:12. [PMID: 20128888]. https://doi.org/10.1186/1741-7015-8-12

Fanjul-Fernández M, Folgueras AR, Cabrera S et al (2010) Matrix metalloproteinases: evolution, gene regulation and functional analysis in mouse models. J Biochim Biophys Acta 1803 (1):3–19

Feng CW, Wang LD, Jiao LH, Liu B, Zheng S, Xie XJ (2000) Expression of p53, inducible nitric oxide synthase and vascular endothelial growth factor in gastric precancerous and cancerous lesions: correlation with clinical features. BMC Cancer 29:2–8

Ferlay JSI, Ervik M, Dikshit R, et al (2013) Cancer incidence and mortality worldwide: IARC CancerBase No. 11; GLOBOCAN 2012 v1.0 2013. Lyon: International Agency for Research on Cancer. http://globocan.iarc.fr

Ferlay J, Soerjomataram I, Dikshit R, Eser S, Mathers C, Rebelo M et al (2015) Cancer incidence and mortality worldwide: sources, methods and major patterns in GLOBOCAN 2012. Int J Cancer 136(5):E359–E386. https://doi.org/10.1002/ijc.29210. PMID: 25220842

Sim Edited by Fiona, McKee Martin (2011) Issues in public health (2nd ed.) Maidenhead: Open University Press. p 74. isbn 9780335244225

Fock KM, Ang TL (2010) Epidemiology of Helicobacter pylori infection and gastric cancer in Asia. J Gastroenterol Hepatol 25:479–486

Fu YP, Kohaar I, Rothman N, Earl J, Figueroa JD, Ye Y (2012) Common genetic variants in the PSCA gene influence gene expression and bladder cancer risk. Proc Natl Acad Sci U S A 109:4974–4979

Fukumura D, Xavier R, Sugiura T et al (1998) Tumor induction of VEGF promoter activity in stromal cells. Cell 94(6):715–725

Garber JE, Offit K (2005) Hereditary cancer predisposition syndromes. J Clin Oncol 23:276–292

Garcia de Herreros A, Baulida J (2012) Cooperation, amplification, and feed-back in epithelial-mesenchymal transition. Biochim Biophys Acta 1825:223–228

Gatti LL, Burbano RR, de Assumpcao PP, Smith Mde A, Payao SL (2004) Interleukin-1beta polymorphisms, *Helicobacter pylori* infection in individuals from northern Brazil with gastric adenocarcinoma. Clin Exp Med 4:93–98

González CA, Agudo A (2012) Carcinogenesis, prevention and early detection of gastric cancer: where we are and where we should go. Int J Cancer 130:745–753

Gonzalez CA, Sala N, Capella G (2002) Genetic susceptibility and gastric cancer risk. Int J Cancer 100:249–260

Gottardi CJ, Wong E, Gumbiner BM (2001) E-cadherin suppresses cellular transformation by inhibiting beta-catenin signaling in an adhesion independent manner. J Cell Biol 153:1049–1060

Grady WM, Markowitz SD (2002) Genetic and epigenetic alterations in colon cancer. Annu Rev Genomics Hum Genet 3:101–128

Guan X, Zhao H, Niu J, Tang D, Ajani JA, Wei Q (2009) The VEGF -634G>C promoter polymorphism is associated with risk of gastric cancer. BMC Gastroenterol 9:77

Gumbiner BM, Mccrea PD (1993) Catenins as mediators of the cytoplasmic functions of cadherins. J Cell Sci Suppl 17:155–158

Hansford S, Kaurah P, Li-Chang H, Woo M, Senz J, Pinheiro H, Schrader KA, Schaeffer DF, Shumansky K, Zogopoulos G, Almeida Santos T, Claro I, Carvalho J, Nielsen C, Padilla S, Lum A, Talhouk A, Baker-Lange K, Richardson S, Lewis I, Lindor NM, Pennell E, MacMillan A, Fernandez B, Keller G, Lynch H, Shah SP, Guilford P, Gallinger S, Corso G, Roviello F, Caldas C, Oliveria C, Pharoah PDP, Huntsman DG (2015) Hereditary diffuse gastric cancer syndrome: CDH1 mutations and beyond. JAMA Oncol 1(1):23–32

Hao B, Miao X, Li Y et al (2006) A novel T-77C polymorphism in DNA repair gene XRCC1 contributes to diminished promoter activity and increased risk of non-small cell lung cancer. Oncogene 25:3613–3620

Heasman SJ, Ridley AJ (2008) Mammalian rho GTPases: new insights into their functions from in vivo studies. Nat Rev Mol Cell Biol 9:690–701

Hruska M, Keefe J, Wert D, Tekinay AB, Hulce JJ, Ibanez-Tallon I (2009) Prostate stem cell antigen is an endogenous lynx1-like prototoxin that antagonizes alpha7-containing nicotinic receptors and prevents programmed cell death of parasympathetic neurons. J Neurosci 29:14847–14854

Huang WY, Chow WH, Rothman N, Lissowska J, Llaca V, Yeager M, Zatonski W, Hayes RB (2005) Selected DNA repair polymorphisms and gastric cancer in Poland. Carcinogenesis 26:1354–1359

Hunter T, Cooper JA (1985) Protein-tyrosine kinases. Annu Rev Biochem 54:897–930

Hyoungseok J, Lim B, Kim M, Kim YS, Kim WH, Ihm C, Noh S-M, Han DS, Yu H-J, Choi BY, Kang C (2009) A regulatory polymorphism at position −309 in PTPRCAP is associated with susceptibility to diffuse-type gastric cancer and gene expression. Neoplasia 11:1340–1347

Jahanarah K, Rai RP, Prasad KN (2016) Role of Helicobacter pylori in gastric cancer: updates. World J Gastrointest Oncol 8(2):147–158

Jasperson KW, Tuohy TM, Neklason DW et al (2010) Hereditary and familial colon cancer. Gastroenterology 138:2044–2058

Jemal A, Bray F, Center MM et al (2011) Global cancer statistics. CA Cancer J Clin 61:69–90

Jones MK, Kawanaka H, Baatar D, Szabo IL, Tsugawa K, Pai R, Koh GY, Kim I, Sarfeh IJ, Tarnawski AS (2001) Gene therapy for gastric ulcers with single local injection of naked DNA encoding VEGF and angiopoietin-1. Gastroenterology 121:1040–1047

Kang GH et al (2008) DNA methylation profiles of gastric carcinoma characterized by quantitative DNA methylation analysis. Lab Invest 88:161–170

Karin M, Greten FR (2005) NF-kappaB: linking inflammation and immunity to cancer development and progression. Nat Rev Immunol 5:749–759

Ke Q, Liang J, Wang LN et al (2008) Potentially functional polymorphisms of the vascular endothelial growth factor gene and risk of gastric cancer. Mol Carcinog 47:647–651

Kelly F, Murray S, Schork N, Topol E (2009) Frazer KA, Murray SS, Schork NJ, Topol EJ. Human genetic variation and its contribution to complex traits. Nat Rev Genet 10: 241-251. Nature reviews. Genetics 10:241–251. https://doi.org/10.1038/nrg2554

Kirsch M, Morz M, Pinzer T, Schackert HK, Schackert G (2009) Frequent loss of the CDKN2C (p18INK4c) gene product in pituitary adenomas. Genes Chromosomes Cancer 48:143–154

Kuphal S, Poser I, Jobin C, Hellerbrand C, Bosserhoff AK (2004) Loss of Ecadherin leads to upregulation of NFkappaB activity in malignant melanoma. Oncogene 23:8509–8519

Lee YY, Derakhshan MH (2013) Environmental and lifestyle risk factors of gastric cancer. Arch Iran Med 16:358–365

Li E (2002) Chromatin modification and epigenetic reprogramming in mammalian development. Nat Rev Genet 3:662–673

Liacini A, Sylvester J, Li WQ et al (2002) Inhibition of interleukin- 1-stimulated MAP kinases, activating protein-1 (AP-1) and nuclear factor kappa B (NF-kappa B) transcription factors downregulates matrix metalloproteinase gene expression in articular chondrocytes. J Matrix Biol 21(3):251–262

Liu Y, Oakeley EJ, Sun L, Jost JP (1998) Multiple domains are involved in the targeting of the mouse DNA methyltransferase to the DNA replication foci. Nucleic Acids Res 26:1038–1045

Lozano R, Naghavi M, Foreman K et al (2012) Global and regional mortality from 235 causes of death for 20 age groups in 1990 and 2010: a systematic analysis for Global Burden of Disease Study 2010. Lancet 380(9859):2095–2128

Lu Y, Chen J, Ding Y et al (2010) Genetic variation of PSCA gene is associated with the risk of both diffuse and intestinal-type gastric cancer in a Chinese population. Int J Cancer 127 (9):2183–2189

Ma J, Sawai H, Ochi N et al (2009) PTEN regulates angiogenesis through PI3K/Akt/VEGF signaling pathway in human pancreatic cancer cells. J Mol Cell Biochem 331(1):161–171

Maeda K, Kang SM, Onoda N, Ogawa M, Kato Y, Sawada T, Chung KH (1999) Vascular endothelial growth factor expression in preoperative biopsy specimens correlates with disease recurrence in patients with early gastric carcinoma. Cancer 86:566–571

Marintchev A, Mullen MA, Maciejewski MW, Pan B, Gryk MR, Mullen GP (1999) Solution structure of the single-strand break repair protein XRCC1 *N*-terminal domain. Nat Struct Biol 6:884–893

Mateus AR, Seruca R, Machado JC et al (2007) EGFR regulates RhoA-GTP dependent cell motility in E-cadherin mutant cells. Hum Mol Genet 16(13):1639–1647

Matsuda A, Motoya S, Kimura S, McInnis R, Maizel AL, Takeda A (1998) Disruption of lymphocyte function and signaling in CD45-associated proteinnull mice. J Exp Med 187:1863–1870

Matsuo YSH, Funahashi H, Takahashi H et al (2004) Enhanced angiogenesis due to inflammatory cytokines from pancreatic cancer cell lines and relation to metastatic potential. J Pancreas 28 (3):344–352

McLean MH, El-Omar EM (2014) Genetics of gastric cancer. Nat Rev Gastroenterol Hepatol 11:664–674

Milne AN, Carneiro F, O'Morain C, Offerhaus GJ (2009) Nature meets nurture: molecular genetics of gastric cancer. Hum Genet 126:615–628

Moon RT, Kohn AD, De Ferrari GV, Kaykas A (2004) Nat Rev Genet 5:691–701

Motoya S, Kitamura K, Matsuda A, Maizel AL, Yamamoto H, Takeda A (1999) Interaction between CD45-AP and protein-tyrosine kinases involved in T cell receptor signaling. J Biol Chem 274:14392–14399

Nagini S (2012) Carcinoma of the stomach: a review of epidemiology, pathogenesis, molecular genetics and chemoprevention. World J Gastrointest Oncol 4:156–169

Nath S, Mukherjee P (2014) MUC1: a multifaceted oncoprotein with a key role in cancer progression. Trends Mol Med 20:332–342

Neureiter D, Herold C, Ocker M (2006) Gastrointestinal cancer—only a deregulation of stem cell differentiation? (review). Int J Mol Med 17:483–489

Olayioye MA (2001) Update on HER-2 as a target for cancer therapy: intracellular signaling pathways of ErbB2/HER-2 and family members. Breast Cancer Res 3:385–389

Oliveira C, Seruca R, Carneiro F (2006) Genetics, pathology, and clinics of familial gastric cancer. Int J Surg Pathol 14:21–33

Pan Y, Bi F, Liu N et al (2004) Expression of seven main Rho family members in gastric carcinoma. Biochem Biophys Res Commun 315:686–691

Park HJ, Yu E, Shim YH (2006) DNA methyltransferase expression and DNA hypermethylation in human hepatocellular carcinoma. Cancer Lett 233:271–278

Parkin DM, Whelan SL, Ferlay J, Raymond L, Young J, editors (1997) Cancer incidence in five continents, volume VII. IARC Scientific Publications No. 143. Lyon: International Agency for Research on Cancer

Parkin DM, Bray F, Ferlay J et al (2005) Global cancer statistics, 2002. CA Cancer J Clin 55 (2):74–108

Pilpilidis I, Kountouras J, Zavos C, Katsinelos P (2011) Upper gastrointestinal carcinogenesis: H. pylori and stem cell cross-talk. J Surg Res 166:255–264

Pinheiro H, Bordeira-Carrico R, Seixas S, Carvalho J, Senz J, Oliveira P et al (2010) Allele-specific CDH1 downregulation and hereditary diffuse gastric cancer. Hum Mol Genet 19:943–952

Pisani P, Bray F, Parkin DM (2002) Estimates of the world-wide prevalence of cancer for 25 sites in the adult population. Int J Cancer 97:72–81

Rahim F, Hajizamani S, Mortaz E et al (2014) Molecular regulation of bone marrow metastasis in prostate and breast cancer. J Bone Marrow Res 2014:405920

Ramamoorti N, Kumarvelu J, Shanmugasundaram GK, Rani K, Banerjea AC (2001) High frequency of G to A transition mutation in the *stromal cell derived factor-1* gene in India, a chemokine that blocks HIV-1 (X4) infection: multiple proteins bind to 3′-untranslated region of SDF-1 RNA. Genes Immunol 2:408–410

Reiter RE, Gu Z, Watabe T, Thomas G, Szigeti K, Davis E, Wahl M, Nisitani S, Yamashiro J, Le Beau MM, Loda M, Witte ON (1998) Prostate stem cell antigen: a cell surface marker overexpressed in prostate cancer. Proc Natl Acad Sci U S A 95:1735–1740

Renner W, Kotschan S, Hoffmann C, Obermayer-Pietsch B, Pilger E (2000) A common 936 C/T mutation in the gene for vascular endothelial growth factor is associated with vascular endothelial growth factor plasma levels. J Vasc Res 37:443–448

Robertson KD, Uzvolgyi E, Liang G, Talmadge C, Sumegi J, Gonzales FA et al (1999) The human DNA methyltransferases (DNMTs) 1, 3a and 3b: coordinate mRNA expression in normal tissues and overexpression in tumors. Nucl Acids Res 27:2291–2298

Ruddon RW (2007). Cancer biology (4th ed.) Oxford University Press. Oxford p. 223. isbn 9780195431

Rugge M, Capelle LG, Fassan M (2014) Individual risk stratification of gastric cancer: evolving concepts and their impact on clinical practice. Best Pract Res Clin Gastroenterol 28:1043–1053

Saeki N, Saito A, Choi IJ, Matsuo K, Ohnami S, Totsuka H, Chiku S, Kuchiba A, Lee YS, Yoon KA et al (2011) A functional single nucleotide polymorphism in mucin 1, at chromosome 1q22, determines susceptibility to diffuse-type gastric cancer. Gastroenterology 140:892–902

Sakamoto H, Yoshimura K, Saeki N et al (2008a) Genetic variation in PSCA is associated with susceptibility to diffuse-type gastric cancer. Nat Genet 40:730–740

Sakamoto H, Yoshimura K, Saeki N, Katai H, Shimoda T, Matsuno Y, Saito D, Sugimura H, Tanioka F, Kato S, Matsukura N, Matsuda N, Nakamura T, Hyodo I, Nishina T, Yasui W, Hirose H, Hayashi M, Toshiro E, Ohnami S, Sekine A, Sato Y, Totsuka H, Ando M, Takemura R, Takahashi Y, Ohdaira M, Aoki K, Honmyo I, Chiku S, Aoyagi K, Sasaki H, Ohnami S, Yanagihara K, Yoon KA, Kook MC, Lee YS, Park SR, Kim CG, Choi IJ, Yoshida T, Nakamura Y, Hirohashi S (2008b) Genetic variation in PSCA is associated with susceptibility to diffuse-type gastric cancer. Nat Genet 40:730–740

Sandra GJ (2001) MUC1, the Renaissance molecule. J Mammary Gland Biol Neoplasia 6:339–353

Senger DR, Galli SJ, Dvorak AM, Perruzzi CA, Harvey VS, Dvorak HF (1983) Tumor cells secrete a vascular permeability factor that promotes accumulation of ascites fluid. Science 219 (4587):983–985

Sharom FJ, Radeva G (2004) GPI-anchored protein cleavage in the regulation of transmembrane signals. Subcell Biochem 37:285–315

Sherr CJ, Roberts JM (1995) Inhibitors of mammalian G1 cyclin-dependent kinases. Genes Dev 9:1149–1163

Solanas G, Porta-de-la-Riva M, Agustí C, Casagolda D, Sánchez-Aguilera F, Larriba MJ, Pons F, Peiró S, Escrivà M, Muñoz A, Duñach M, de Herreros AG, Baulida J (2008) J Cell Sci 121:2224–2234

Staal FJ, Luis TC, Tiemessen MM (2008) WNT signalling in the immune system: WNT is spreading its wings. Nat Rev Immunol 8:581–593

Stemmler MP (2008) Cadherins in development and cancer. Mol BioSyst 4:835–850

Summy JM, Gallick GE (2003) Src family kinases in tumor progression and metastasis. Cancer Metastasis Rev 22:337–358

Sun Y et al (2004) Methylation of p16 CpG islands associated with malignant transformation of gastric dysplasia in a population-based study. Clin Cancer Res 10:5087–5093

Suriano G, Mulholland D, de Wever O, Ferreira P, Mateus AR, Bruyneel E, Nelson C, Mareel M, Yokota J, Huntsman D et al (2003) The intracellular E-cadherin germline mutation V832M lacks the ability to mediate cell–cell adhesion and to suppress invasion. Oncogene 22:5716–5719

Tahara T, Shibata T, Nakamura M et al (2009) Effect of polymorphisms in the 3¢untranslated region (3′-UTR) of vascular endothelial growth factor gene on gastric cancer and peptic ulcer diseases in Japan. Mol Carcinog 48:1030–1037

Takeda A, Matsuda A, Paul RM, Yaseen NR (2004) CD45-associated protein inhibits CD45 dimerization and up-regulates its protein tyrosine phosphatase activity. Blood 103:3440–3447

Takeichi M (1995) Morphogenetic roles of classic cadherins. Curr Opin Cell Biol 7:619–627

Valastyan S, Weinberg RA (2011) Tumor metastasis: molecular insights and evolving paradigms. Cell 147:275–292

Wang J, El-Bahrawy M (2015) Expression profile of mucins (MUC1, MUC2, MUC5AC, and MUC6) in ovarian mucinous tumours: changes in expression from benign to malignant tumours. Histopathology 66:529–535

Wei ZW, Xia GK, Wu Y et al (2015) CXCL1 promotes tumor growth through VEGF pathway activation and is associated with inferior survival in gastric cancer. J Cancer Lett 359(2):335

World Cancer Report 2014

World Cancer Report 2014. World Health Organization. 2014. pp. Chapter 5.4. isbn 9283204298

Xie WQ, Tan SY, Wang XF (2014) MiR-146a rs2910164 polymorphism increases risk of gastric cancer: a meta-analysis. World J Gastroenterol 20(41):15440–15447. https://doi.org/10.3748/wjg.V20.I41.15440

Yagi T, Takeichi M (2000) Cadherin superfamily genes: functions, genomic organization, and neurologic diversity. Genes Dev 14:1169–1180

Yamamoto S, Yasui W, Kitadai Y et al (1998) Expression of vascular endothelial growth factor in human gastric carcinomas. Pathol Int 48(7):499–506

Yancopoulos GD, Davis S, Gale NW, Rudge JS, Wiegand SJ, Holash J (2000) Vascular-specific growth factors and blood vessel formation. Nature 407(6801):242–248

Yin M, Hu Z, Tan D, Ajani JA, Wei Q (2009) Molecular epidemiology of genetic susceptibility to gastric cancer: focus on single nucleotide polymorphisms in gastric carcinogenesis. Am J Transl Res 1:44–54

Yoon H, Kim N (2015) Diagnosis and management of high-risk group for gastric cancer. Gut Liver 9:5–17

Yoshio Y, Mototsugu K, Masahiro A (2008) Geographic differences in gastric cancer incidence can be explained by differences between Helicobacter pylori strains. Intern Med 47(12):1077–1083

Yurgelun MB, Boland CR (2017) New cancer genes and inherited colorectal cancer risk: caveat emptor. Gastroenterology 152(1):12–13. https://doi.org/10.1053/j.gastro.2016.11.027

Zali H, Rezaei-Tavirani M, Azodi M (2011) Gastric cancer: prevention, risk factors and treatment. Gastroenterol Hepatol 4:175–185

Zhang S et al (2009) RhoA regulates G1-S progression of gastric cancer cells by modulation of multiple INK4 family tumor suppressors. Mol Cancer Res 7:570–580

Zhifang J, Wu X, Cao D, Wang C, You L, Jin M, Wen S, Cao X, Jiang J (2016) Polymorphisms of the DNA methyltransferase 1 gene predict survival of gastric cancer patients receiving tumorectomy. Hindawi Publishing Corporation, disease markers. 2016:9. https://doi.org/10.1155/2016/8578064

Chapter 8
Overview of Early Detection of Gastrointestinal Cancer

Pola Sudhakar, Pavani Sanapala, and B. Pratap Naidu

Abstract Considering both incidence and mortality rates of gastric cancer (GC) being low affected, the disease remains a frequent source of cancer deaths globally. The prediction of GC is based on its staging, hence detecting at an early stage is crucial for a long life that diagnosed at the later. Identifying the cause and treating it will help the patients to sustain a better prognosis. Both genetic and non-genetic factors play an essential role in causing GC. Besides, viral infection by *Helicobacter pylori* has been proven to cause GC. Identifying and characterizing molecular biomarkers, epigenetic alterations, long non-coding RNAs, circulating tumor DNA and RNA, abnormal methylation with the help of advanced techniques such as microarray profiling, high-throughput techniques, endoscopy, screening body fluids, quantitative PCR, and the advanced next-generation sequencing increase the source of detecting and identifying the gastric cancer at the earliest. However, certain drugs have been administrated to treat early gastric cancer. This chapter reviews in detail the information regarding prognostic and noninvasive biomarkers for the early detection of gastric cancer.

Keywords Gastric cancer · Biomarkers · miRNAs · Circulating tumor RNA · Metabolic biomarkers

P. Sudhakar (✉) · P. Sanapala
Department of Biotechnology, Andhra University, Visakhapatnam, Andhra Pradesh, India
e-mail: sudhakar@andhrauniversity.edu.in

B. P. Naidu
Department of Biotechnology, Krishna University, Machilipatnam, Andhra Pradesh, India

P. Veera Bramhachari, N. R. R. Neelapu (eds.), *Recent Advancements in Biomarkers and Early Detection of Gastrointestinal Cancers*, Diagnostics and Therapeutic Advances in GI Malignancies, https://doi.org/10.1007/978-981-15-4431-6_8

8.1 Introduction

Gastrointestinal cancer (GC) is rated the fifth most common cancer diagnosed and the third principal cause of mortalities among all cancers globally (Bray et al. 2018). Over the past 10 years, the incidence of GC has publicized tremendous reduction, but the five-year survival index showed patients of GC progression to advance stage from early stages (Luo and Li 2019). Treating cancer on an early-onset might reduce the risk and would be a relief to the global burden of the disease so, early detection of any cancer increases the chance of survival on successful prognosis and treatment. Usually, diagnosis or detection of cancer involves two primary mechanisms: (1) education to prop up early diagnosis, and (2) screening. Many programs have been instituted for better cancer outcomes, one such program established in the year 2010 was Be Clear on Cancer (BCOC). Diagnosis using different techniques like endoscopy and biopsy have been in force, but these techniques lack detecting the disease at the earliest, whereas screening for molecular markers aid in revealing early gastric cancer (EGC).

Research on cancer by identifying and characterizing molecular biomarkers, tumor markers, and genetic alteration such as bulky addition or loss of chromosomal, single nucleotide polymorphisms, epigenetics, mutational alteration, histone protein modification, abnormal DNA methylation, over-expression of miRNA, circulating tumor RNA and DNA, and lncRNAs (long non-coding RNAs) has been on forth to detect cancer at the earliest (Cancer Genome Atlas Research Network and Analysis Working Group 2014). In recent times, the application of high-throughput technologies has taken new approaches into molecular pathogenesis, ensuing a novel classification of gastric carcinoma in support of their genomic characterization. Cancer genome atlas has classified GC into four subtypes basing on the targeted material: (a) Epstein–Barr virus-infected tumors, (b) Microsatellite instability tumors, (c) Genomically stable tumor, and (d) Chromosomally unstable cancer (Cristescu et al. 2015). Another new classification was given by the Asian Cancer Research Group, which is microsatellite stable and instability cancer (Patel et al. 2017).

Helicobacter pylori, a Gram-negative, microaerophilic bacterium commonly resite in the human stomach. Recent studies confirmed the risk of gastric carcinoma by *H. pylori* infection that causes mild to severe gastritis that carries on for a lifetime if not treated with antimicrobial drugs (Correa 1995). The anticipation of dietary intake and screenings reduces the risk of gastric cancer. However, intervention studies such as annihilation through chemoprevention trials have revealed possible strategies. So, understanding the interrelationship of disease and the factors can help researchers and scientists to drive towards novel approaches in the field of reducing the disease progression (Yoon et al. 2011). This chapter aims to discuss the recent advances for early diagnosis of gastric cancer (Fig. 8.1).

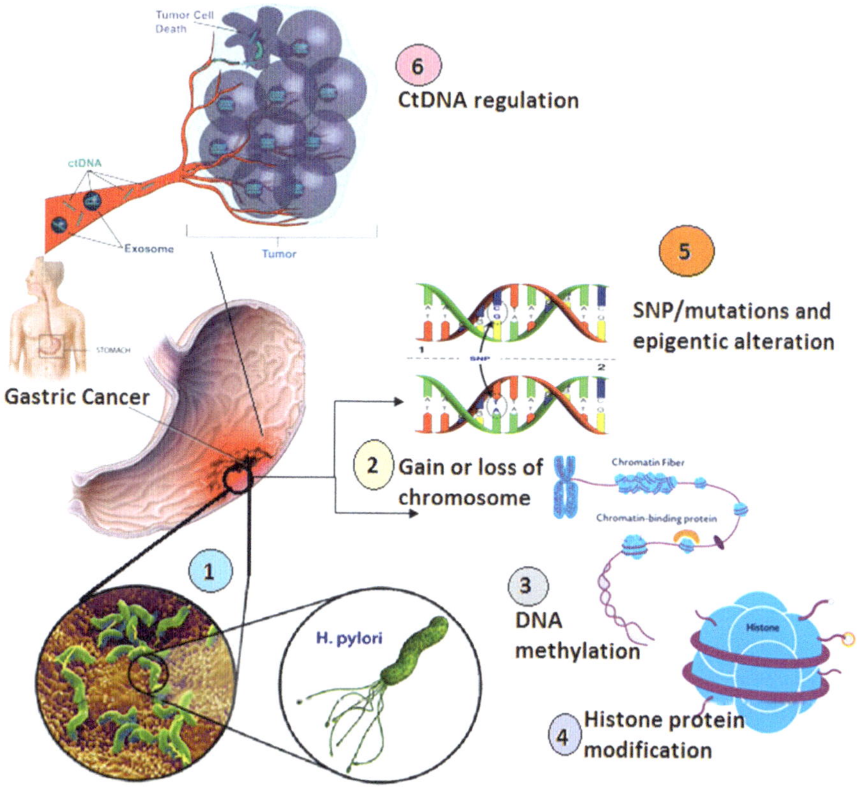

Fig. 8.1 Advance process in detecting gastric cancer

8.2 Screening of GC at the Earliest

Gastroscopy is money-making for the gastroenterologist nowadays; however, the technique is not affordable by the patient since it is expensive. Besides this drawback, it also holds a few complications. Screening procedures that are economically priced, noninvasive and apt for the general population have been required (Tan and Fielding 2006). The progress of high-tech techniques results in molecular markers capable of identifying the disease at the earliest, calculate the disease outcome, and aid admittance for proper therapy.

8.2.1 Possible Metabolic Biomarkers for GC Metastasis

Metastasis, in general, spreads disease from one organ to another, either adjacent or distant organs. Pathophysiology of the disease confirms the deaths of gastric cancer

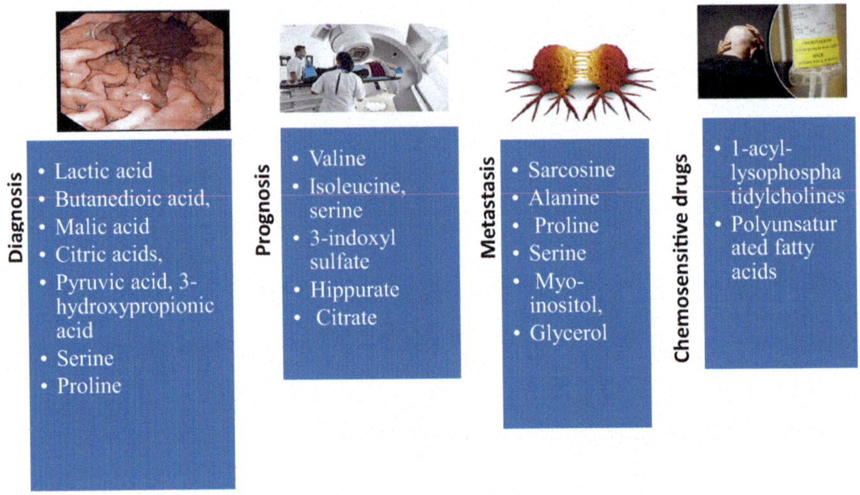

Fig. 8.2 Metabolite markers in connection with detection and treatment

are primarily a result of metastasis, which can identify metabolic markers. Studies on the metabolomics of human engraft models illustrated principal mechanisms of GC metastasis and probable biomarkers for early diagnosis (Chen et al. 2010). Of the 30 metabolites identified, glutamine was the major metabolite that showed 1.71-fold diminution in expression in the metastatic group rather than non-metastatic. Likewise, praline the upregulated metabolite showed 2.45 times increased expression (Chen et al. 2010). Few studies reported significant variation between the cancerous and non-cancerous group for metabolite composition. Metabolite such as proline, leucine, serine, malic, and lactic acid are known to play an essential role in metastasis of GC (Fig. 8.2). The biomarkers derivative of these metabolites pave pathways that can be support for treating GC at the onset (Hu et al. 2011). Many techniques, namely NMR spectroscopy, liquid and gas chromatography (GC), GC-MS (mass spectrophotometer), capillary electrophoresis-MS, and Fourier spectroscopy are widely used in metabolomics analysis (Jayavelu and Bar 2014).

8.2.2 *Acid Suppression Therapy*

Usually, the symptoms EGC are impossible to differentiate compared to benign; such patients are treated with acid suppression drugs, as well as H^+ pump inhibitors or hydrogen blockers before gastroscopy. Comparative studies revealed H^+ pump inhibitors were healing malignant stomach ulcers with 4 weeks of onset, whereas biopsy and endoscopy are required after acid suppression (Corrigan et al. 1997; Taylor et al. 1978; Wayman et al. 2000).

8.2.3 Endoscopy Techniques

Modern advances in endoscopic tools have enhanced the sensitivity of identifying EGC. The techniques follow three main steps: (a) detecting suspicious lesions, (b) characterizing the lesions, and finally (c) an accurate diagnosis. Different techniques, namely magnifying and chromoendoscopy, modern high-resolution virtual chromoendoscopy, confocal laser end microscopy, and flexible spectral imaging, have been in use.

Chromoendoscopy, in combination with indigo carmine, is able for recognition and handling targeted biopsies of abnormal areas of the gastric mucosa so that the dye augments tissue abnormality through a high magnification image (Sasako 1997). Magnifying endoscopy assesses gastric lesions at microvascular construction that provides an option of envisaging the histological nature of cancer. Conversely, magnifying endoscopy does not investigate the whole of the gastric mucosa. Another novel technique using infrared light gives deep tissue incursion using an infrared video endoscope (Mataki et al. 2003). Likewise, light-induced fluorescence endoscopy, equipment-based image enhanced endoscopy and endoscopy ultrasonography gave promising results. Endoscopic test plays a vital role in early detection of cancer but the accuracy of detection primarily depends on the endoscopist so that they have a pure knowledge to determine lesions. Combining to or more techniques may increase the scope of better diagnosis so developing minimal invasive endoscopic methods is a challenge to the researchers.

8.2.4 Serological Test

Quite a few tests using blood samples have been examined to institute aptness as screening tools to detect patients with GC. The tests include screening for pepsinogen, gastrin 17, and *Helicobacter pylori* antibody. Pepsinogen, a pepsin precursor, be present in two forms, pepsinogen I (PG I) and II (PG II). The difference between the two is their secretion. PG I is secreted mainly by corpus cells where PG II is secreted in cells of the antrum, corpus, Brunner's gland of the duodenum. The ratio of PG I and PG II is concentrated due to the overproduction of PG II in the antrum, duodenum, and corpus cells (Kikuchi et al. 2000). A study by Kitahara and colleagues revealed the significance of pepsinogen screening is competent in diagnosing GC in patients of atrophic gastritis. The process includes a mixture of PG I and PG I/PG II ratio as an endpoint. However, the test does not apply to mild atrophic gastritis (Kitahara et al. 1999).

As already discussed, *H. pylori* bacteria play a role in infecting the individuals causing gastric cancer. The antibody of this species acts as a marker in screening dystrophy in patients below 45 years. The test showed a 97% sensitivity and 87% specificity for GC (Sobala et al. 1991). Dissimilarity findings by Whiting and Co. reported reduced activity of *H. pylori* antibody in patients above 40 years and over 30% had missed diagnosis (Whiting et al. 1998). A study in the year 2005, screening 9293 healthy Japanese with pepsinogen and *H. pylori* antibody showed promising results and concluded the duo be exclusive screening biomarkers (Watabe et al. 2005).

Gastrin 17, a form of gastrin secreted in G cells, was found in the antrum. In patients of atrophic gastritis, the thrashing of antral G cells resulted in reduced gastrin levels. G-17 subsequently screened these levels as an indicator (Sipponen et al. 1990). Research findings by Sipponen et al. confirmed the use of serology biomarkers, namely pepsinogen, gastrin, and *H. pylori* antibody together efficiently detected diverse models of gastritis with sensitivity and specificity percentage of 89% and 93%, correspondingly.

8.2.5 Tumor Biomarkers

For early diagnosis and detection of gastric cancer clinically, tumor marks such as carcinoembryonic antigen (CEA), carbohydrate antigen (CA), and alpha-fetoprotein (AFP) were used (Tsai et al. 2016a). However, the specificity and sensitivity of these biomarkers are stated to be reduced (Tsai et al. 2016a; Tong et al. 2016). Many study evidence explained the improved expression level of oncogenes in gastric cancer where they excite cell cycle and tumor cells growth and also by inhibiting apoptosis. Several genes were identified that showed a positive response in identifying gastric cancer initially. The genes, namely xeroderma pigmentosum group (*xpg*), *stc1*, *iftim1* (interferon-induced transmembrane protein 1), matrix metalloproteinase 9 (*MMP9*), and pituitary tumor transforming gene 1 were known to determine GC (Kanda and Kodera 2015).

xpg/ercc5 (xeroderma pigmentosum group G/excision repair cross-complementing group 5) an enzyme from the nucleotide excision repair system is said to involve in revamping DNA lesions that are a result of genomic instability. The expression of *ercc5* was significantly reported to have progressed results towards GC gastritis and also coupled with tumor development (Deng et al. 2014). By microarray profiling, the gene *ifitm1* was detected in upregulating tumor cell lines and gastric cancer tissues. The gene *ifitm1* also plays a vital role in implicating invasions and transfer GC cells to enhance inflammatory response that is a part of tumor progression (Lee et al. 2012). The gene MMP-9 an enzyme plays a role in tumor growth expansion, metastasis, and invasion in gastric carcinoma (Zheng et al. 2006).

Microarray profiling discovered few overexpressed genes such as *KRT17, COL10A1, KIAA1199, SPP1, IL11, S100A2*, and *MMP3* that are related to tumor progression. Out of these, the more used candidate markers were *KRT17 and COL10A1* that had enhanced expression for EGC (Chivu et al. 2010). Likely, tumor suppressor genes were also used for early detection as they presented reduced expression in GC patients that resulted in hastened cell growth, declined inhibition of oncogene expression, and the development of cell growth.

A remarkable biomarker gastrokine 1 (*GKN1*) is extensively expressed in the surface lumen of gastric tissue that indulges in upholding mucosal integrity in the stomach but is not found in GC (Altieri et al. 2017). The gene also acts as a tumor suppressor and modulates apoptosis signaling in GC. These factors and it is lower expression consider the gene as an indicator of increased risk of gastric carcinogenesis (Watanabe et al. 2009).

8.2.6 Circulating Tumor Cells

Cancer cells freed from a general tumor or the metastatic sites circulate in the blood that is defined as circulating tumor cells (CTCs). The cells are usually sensed by epithelial cell adhesion molecule and cytokeratins. CTCs as diagnostic marker is commonly present in the blood of GC patients.

This gives predictive information after surgical and chemotherapic activities (Haber and Velculescu 2014). CTCs were initially illustrated as expressing epithelial cell markers. *EpCAm*, cytokeratin (CK): *CK8*, *CK18*, and *CK19* are CTCs showing an adverse effect for CD45 (Allard et al. 2004). Studies showed the occurrence of CTCs in circulating tumor microemboli representing poor prognosis and controlling disease progression (Chinen et al. 2017).

The soaring heterogeneity of CTCs provoked researchers to expand various methodologies to augment, isolate, and itemize them basing on specific phenotypic and molecular characterization. In general, there are two methodologies used for the isolation and enumeration of CTCs. One is the biological method CellSearch platform for enumeration and the other be physical method, namely Food and Drug Administration (FDA) for clinical purposes. These methods detect *EpCAm*, *CK8*, *CK18*, and *CK19 excluding CD45*. Using cell-size and phenotype-based systems, as centrifugal microfluidic system based on fluid-assisted separation technique (FAST), or Cascaded Inertial Focusing Microfluidic device, coupled with detection of an extended panel of markers might identify a different subpopulation of CTCs with higher efficiency (Kang et al. 2017; Abdulla et al. 2018). Another technique, immunostaining-fluorescence in situ hybridization (iFISH) platform, claimed to be more sensitive than the CellSearch™ to detect and characterize CTCs in advanced GC patients.

8.2.7 Circulating Tumor DNA (CtDNA) as a Biomarker

CtDNA investigation developed the liquid biopsy to detect traces of tumor molecular moving body fluids and confer a deeper approach into the cancer heterogeneity, early detection of biological markers, finding therapeutic agents, instances assessment of healing response, and potential resistance and prediction. In general, ctDNA corresponds to only parts of the cell-free circulating DNA (cfDNA) that is noticeably amplified in an advanced stage of the disease (Bettegowda et al. 2014). Studies also showed traces of ctDNA in plasma samples of EGC patients (Alix-Panabières and Pantel 2016; Sumbal et al. 2018). The levels of ctDNA were interrelated with that of vascular invasions, peritoneal repetition, and diagnosis (Fang et al. 2016). In EGC patients, competence for diagnosis was reported with cfDNA consisting of *rassfla* and *apc* promoter hypermethylation (Balgkouranidou et al. 2015). Advanced biological techniques such as multiplex MS SNP genotyping, RT quantitative PCR, digital droplet PCR, next-generation sequencing, and advanced nuclear quantification technology were in use to analyze ctDNA in GC individuals to detect the disease at the most basic (Shoda et al. 2015, 2017; Kato et al. 2018).

8.2.8 CircRNAs

Circular RNAs (CircRNAs) the latest division of non-coding RNA that appears as a closed-loop without ends 5′ and 3′ (Memczak et al. 2013). The presence of CircRNAs in RNA virus was initially reported but, studies found stable and preserved CircRNAs sequence in almost all eukaryotes that organize gene expression by miRNAs connection through microarray profiling and high-throughput RNA sequencing (Chen 2016). The role of CircRNAs was noted in many diseases especially as tumor growth and metastasis (Li et al. 2015). Several CircRNAs are discovered and shown to express in gastric tissues. Of all the types, *hsa_circ_0000026*, *hsa_circRNA_400071*, *hsa_circRNA_000543*, and *hsa_circRNA_001959* are reported to have expressed in multiples in GC. *hsa_circ_0000026* explained expression of downregulation whereas the other showed differential gene expression (Sui et al. 2017; Huang et al. 2017). Few genes of CirRNA namely *cd44*, *cxxc5*, *myh9,* and *malat1*suggested to have a role in growth and tumorigenesis. The overall findings discussed here unlocked the path for plasma circRNA profiling that aims to detect definite diagnostic and prognostic circular RNA markers for early gastric cancer individuals.

8.2.9 LncRNAS Transcriptomes Marker

Long non-coding RNA (LncRNA) are transcripts of 200 nucleotides long with nix or partial perspective towards protein-coding. LncRNAS is labeled as a transcriptomes marker due to its regulation in transcription, translation, cellular differentiation, cell cycle processes, and gene expression (Wang et al. 2015). The factor of its high stable condition while moving in body fluids and also their altitude in tumor tissues made the marker useful to diagnose GC patients at the earliest (Shi et al. 2016; Bolha et al. 2017). A study by Cao and his colleagues revealed 88 differential LncRNAs where 71 showed upregulation and 74 downregulation (Cao et al. 2013). Zhou et al. suggested the use of lncRNA and H19 as potential biomarkers to detect and monitor

GC especially for early screening. Besides the discussed markers, *lncRNA PVT1*candidate, *TINCR, CCAT2, AOC4P, BANCR, CUDR, LSINCT-5, PTENP1,* and *LINC00857* also acts as a possible marker for GC diagnosis (Zhou et al. 2015; Yuan et al. 2016; Zhang et al. 2017; Dong et al. 2015).

8.2.10 Epigenetic Alteration-Methylation

Few definite genes such as *p16nk4a, tcf4,* DNA repair (*hmlh1* and *mgmt*), cell growth/differentiation (*hoxd10, hai-2/spint2, ndrg2*), transcriptional regulation (*hltf, pax6, znf545, runx3*), cell adhesion/invasion/migration (*cdh1, cdh4, apc, flnc, lox, timp3, tsp1*), apoptosis (*bnip3, xiap, bnip3, bcl2, cacna2d3, dapk, gpx3, pcdh10, pcdh17, casp8, xaf1*), angiogenesis (*thbs-1* and *p73*), STAT pathway (*socs-1*), Ras pathway (*rassf1a, rassf2, hdab2ip, rkip*), Wnt pathway (*dkk-3, ctnnb1*), in addition to in multidrug resistance genes (*mdr1, gstp1*) are reported to regulate in GC persons (Qu et al. 2013; Kazmi et al. 2018). The association of these gene show varied results such as highly methylated in dysplasia and EGC whereas few show lower methylation in advanced stage (Watanabe et al. 2009).

8.2.11 MiRNAs as a Diagnostic and Prognostic Marker

MicroRNA (miRNA) small non-coding RNA of 19–25 nucleotides long regulates in epigenetic mechanisms such as proliferation, differentiation, cellular processes, and apoptosis. These RNAs are functioned as oncogenes or tumor suppressors basing on the targeted gene (Guimarães et al. 2018). Measuring the serum levels and peripheral blood mononuclear cells show miRNA 21 is overexpressed in gastric patients with a sensitivity of 90%. Few other types of CA199 and CEA reported only 50% specificity (Wu et al. 2015). Gene miR-376c and arid4a are shown to upregulate and downregulate in tissue, plasma, and urine of GC patients (Hung et al. 2017). To date, more than 2500 miRNA genes have been distinguished to express in GC patients. Different studies investigation concluded the presence of various miRNAs, namely miR-196a and 196b, miR-501-3p, miR-143-3p, miR-451a, miR-146a, miR-16, miR-25, miR-92a, miR-451, and miR-486-5p, miR-200a-3p, miR-296-5p, miR-132-3p, miR-485-3p, and miR-22-5p, miR10b-5p, miR132-3p, miR185-5p, and miR195-5p function as noninvasive biomarkers, upregulated and downregulated in gastric cancer persons (Tsai et al. 2016b; Jiang et al. 2017; Zeng et al. 2012; Wang et al. 2018; Zhou et al. 2017).

8.3 Conclusion

Globally, gastric cancer death rate has increased ten times compared to other cancers. The majority of cancer is diagnosed at the final stage as the disease shows no symptoms at the initial stage due to this treating the patient at the earliest is limited. The detection of biomarkers for the disease diagnosis and prognosis aids in curing the disease at the early stage so, studies are directed towards identification and validation of noninvasive markers, cost-effective, highly stable, specific, and sensitive to the GC patients. Few markers mainly ctDNA, ctRNA, lncRNAs, circRNAs, and miRNAs were discovered and reported promising results for early diagnosis of gastric cancer. However, still, strategies have been to plan to get improved and enriched techniques to detect the disease at the earliest.

Acknowledgments The authors are grateful to authorities of Andhra University Visakhapatnam, India, and Krishna University, Machilipatnam, for providing necessary facilities to carry out the research work and for extending constant support.

Conflict of Interest The authors declare that there is no potential conflict of interest.

References

Abdulla A, Liu W, Gholamipour-Shirazi A, Sun J, Ding X (2018) High-throughput isolation of circulating tumor cells using cascaded inertial focusing microfluidic channel. Anal Chem 90 (7):4397–4405

Alix-Panabières C, Pantel K (2016) Clinical applications of circulating tumor cells and circulating tumor DNA as liquid biopsy. Cancer Discov 6(5):479–491

Allard WJ, Matera J, Miller MC, Repollet M, Connelly MC, Rao C et al (2004) Tumor cells circulate in the peripheral blood of all major carcinomas but not in healthy subjects or patients with nonmalignant diseases. Clin Cancer Res 10(20):6897–6904

Altieri F, Di Stadio CS, Federico A, Miselli G, De Palma M, Rippa E, Arcari P (2017) Epigenetic alterations of gastrokine 1 gene expression in gastric cancer. Oncotarget 8(10):16899

Balgkouranidou I, Matthaios D, Karayiannakis A, Bolanaki H, Michailidis P, Xenidis N et al (2015) Prognostic role of APC and RASSF1A promoter methylation status in cell free circulating DNA of operable gastric cancer patients. Mutat Res 778:46–51

Bettegowda C, Sausen M, Leary RJ, Kinde I, Wang Y, Agrawal N, et al (2014) Detection of circulating tumor DNA in early-and late-stage human malignancies. Sci Transl Med 6 (224):224ra24–224ra24

Bolha L, Ravnik-Glavač M, Glavač D (2017) Long noncoding RNAs as biomarkers in cancer. Dis Markers 2017:7243968

Bray F, Ferlay J, Soerjomataram I, Siegel RL, Torre LA, Jemal A (2018) Global cancer statistics 2018: GLOBOCAN estimates of incidence and mortality worldwide for 36 cancers in 185 countries. CA Cancer J Clin 68(6):394–424

Cancer Genome Atlas Research Network, & Analysis Working Group (2014) Comprehensive molecular characterization of gastric adenocarcinoma. Nature 513:202–209

Cao WJ, Wu HL, He BS, Zhang YS, Zhang ZY (2013) Analysis of long non-coding RNA expression profiles in gastric cancer. World J Gastroenterol 19(23):3658

Chen LL (2016) The biogenesis and emerging roles of circular RNAs. Nat Rev Mol Cell Biol 17 (4):205

Chen JL, Tang HQ, Hu JD, Fan J, Hong J, Gu JZ (2010) Metabolomics of gastric cancer metastasis detected by gas chromatography and mass spectrometry. World J Gastroenterol 16(46):5874

Chinen LTD, Abdallah EA, Braun AC, de Campos BDCT, Flores P, Corassa M et al (2017) Circulating tumor cells as cancer biomarkers in the clinic. In: Isolation and molecular characterization of circulating tumor cells. Springer, Cham, pp 1–41

Chivu ME, Necula LG, Dragu D, Badea L, Dima SO, Tudor S et al (2010) Identification of potential biomarkers for early and advanced gastric adenocarcinoma detection. Hepatogastroenterology 57(104):1453–1464

Correa P (1995) Helicobacter pylori and gastric carcinogenesis. Am J Surg Pathol 19:S37–S43

Corrigan NT, O'Riordain MG, McEntee G (1997) Modern anti-ulcer regimens and gastric carcinoma-is there a delay in definitive treatment. Br J Surg Suppl 84:62–63

Cristescu R, Lee J, Nebozhyn M, Kim KM, Ting JC, Wong SS et al (2015) Molecular analysis of gastric cancer identifies subtypes associated with distinct clinical outcomes. Nat Med 21(5):449

Deng N, Liu JW, Sun LP, Xu Q, Duan ZP, Dong NN, Yuan Y (2014) Expression of XPG protein in the development, progression and prognosis of gastric cancer. PLoS One 9(9):e108704

Dong L, Qi P, Xu MD, Ni SJ, Huang D, Xu QH et al (2015) Circulating CUDR, LSINCT-5 and PTENP 1 long noncoding RNA s in sera distinguish patients with gastric cancer from healthy controls. Int J Cancer 137(5):1128–1135

Fang WL, Lan YT, Huang KH, Liu CA, Hung YP, Lin CH et al (2016) Clinical significance of circulating plasma DNA in gastric cancer. Int J Cancer 138(12):2974–2983

Guimarães CTU, Martins NNF, da Silva Oliveira KC, Almeida CM, Pinheiro TM, Gigek CO, Calcagno DQ (2018) Liquid biopsy provides new insights into gastric cancer. Oncotarget 9 (19):15144

Haber DA, Velculescu VE (2014) Blood-based analyses of cancer: circulating tumor cells and circulating tumor DNA. Cancer Discov 4(6):650–661

Hu JD, Tang HQ, Zhang Q, Fan J, Hong J, Gu JZ, Chen JL (2011) Prediction of gastric cancer metastasis through urinary metabolomic investigation using GC/MS. World J Gastroenterol 17 (6):727

Huang YS, Jie N, Zou KJ, Weng Y (2017) Expression profile of circular RNAs in human gastric cancer tissues. Mol Med Rep 16(3):2469–2476

Hung PS, Chen CY, Chen WT, Kuo CY, Fang WL, Huang KH et al (2017) miR-376c promotes carcinogenesis and serves as a plasma marker for gastric carcinoma. PLoS One 12(5):e0177346

Jayavelu ND, Bar NS (2014) Metabolomic studies of human gastric cancer. World J Gastroenterol 20(25):8092

Jiang X, Wang W, Yang Y, Du L, Yang X, Wang L et al (2017) Identification of circulating microRNA signatures as potential noninvasive biomarkers for prediction and prognosis of lymph node metastasis in gastric cancer. Oncotarget 8(39):65132

Kanda M, Kodera Y (2015) Recent advances in the molecular diagnostics of gastric cancer. World J Gastroenterol 21(34):9838

Kang HM, Kim GH, Jeon HK, Kim DH, Jeon TY, Park DY et al (2017) Circulating tumor cells detected by lab-on-a-disc: role in early diagnosis of gastric cancer. PLoS One 12(6):e0180251

Kato S, Okamura R, Baumgartner JM, Patel H, Leichman L, Kelly K et al (2018) Analysis of circulating tumor DNA and clinical correlates in patients with esophageal, gastroesophageal junction, and gastric adenocarcinoma. Clin Cancer Res 24(24):6248–6256

Kazmi HR, Kumari S, Tiwari S, Khanna A, Narayan G (2018) Epigenetic mechanisms and events in gastric cancer-emerging novel biomarkers. Pathol Oncol Res 24(4):757–770

Kikuchi S, Kurosawa M, Sakiyama T, Tenjin H, Miki K, Wada O, Inaba Y (2000) Long-term effect of Helicobacter pylori infection on serum pepsinogens. Jpn J Cancer Res 91(5):471–476

Kitahara F, Kobayashi K, Sato T, Kojima Y, Araki T, Fujino MA (1999) Accuracy of screening for gastric cancer using serum pepsinogen concentrations. Gut 44(5):693–697

Lee J, Goh SH, Song N, Hwang JA, Nam S, Choi IJ et al (2012) Overexpression of IFITM1 has clinicopathologic effects on gastric cancer and is regulated by an epigenetic mechanism. Am J Pathol 181(1):43–52

Li P, Chen S, Chen H, Mo X, Li T, Shao Y et al (2015) Using circular RNA as a novel type of biomarker in the screening of gastric cancer. Clin Chim Acta 444:132–136

Luo M, Li L (2019) Clinical utility of miniprobe endoscopic ultrasonography for prediction of invasion depth of early gastric cancer: a meta-analysis of the diagnostic test from PRISMA guideline. Medicine 98(6):e14430

Mataki N, Nagao S, Kawaguchi A, Matsuzaki K, Miyazaki J, Kitagawa Y et al (2003) Clinical usefulness of a new infrared video endoscope system for diagnosis of early stage gastric cancer. Gastrointest Endosc 57(3):336–342

Memczak S, Jens M, Elefsinioti A, Torti F, Krueger J, Rybak A et al (2013) Circular RNAs are a large class of animal RNAs with regulatory potency. Nature 495(7441):333–338

Patel TN, Roy S, Ravi R (2017) Gastric cancer and related epigenetic alterations. Ecancermedicalscience 11:714

Qu Y, Dang S, Hou P (2013) Gene methylation in gastric cancer. Clin Chim Acta 424:53–65

Sasako M (1997) Radical surgery. Gastric cancer

Shi T, Gao G, Cao Y (2016) Long noncoding RNAs as novel biomarkers have a promising future in cancer diagnostics. Dis Markers 2016:9085195

Shoda K, Masuda K, Ichikawa D, Arita T, Miyakami Y, Watanabe M et al (2015) HER2 amplification detected in the circulating DNA of patients with gastric cancer: a retrospective pilot study. Gastric Cancer 18(4):698–710

Shoda K, Ichikawa D, Fujita Y, Masuda K, Hiramoto H, Hamada J et al (2017) Clinical utility of circulating cell-free Epstein–Barr virus DNA in patients with gastric cancer. Oncotarget 8 (17):28796

Sipponen P, Valle J, Varis K, Kekki M, Ihamäki T, Siurala M (1990) Fasting levels of serum gastrin in different functional and morphologic states of the antrofundal mucosa: an analysis of 860 subjects. Scand J Gastroenterol 25(5):513–519

Sobala GM, Pentith JA, Axon ATR, Dixon MF, Crabtree JE, Rathbone BJ et al (1991) Screening dyspepsia by serology to Helicobacter pylori. Lancet 338(8759):94–96

Sui W, Shi Z, Xue W, Ou M, Zhu Y, Chen J et al (2017) Circular RNA and gene expression profiles in gastric cancer based on microarray chip technology. Oncol Rep 37(3):1804–1814

Sumbal S, Javed A, Afroze B, Zulfiqar HF, Javed F, Noreen S, Ijaz B (2018) Circulating tumor DNA in blood: future genomic biomarkers for cancer detection. Exp Hematol 65:17–28

Tan YK, Fielding JW (2006) Early diagnosis of early gastric cancer. Eur J Gastroenterol Hepatol 18 (8):821–829

Taylor R, Lovell D, Menzies-Gow N, La Brooy SJ, Misiewicz JJ (1978) Misleading response of malignant gastric ulcers to cimetidine. Lancet 311(8066):686–688

Tong W, Ye F, He L, Cui L, Cui M, Hu Y et al (2016) Serum biomarker panels for diagnosis of gastric cancer. Onco Targets Ther 9:2455

Tsai MM, Wang CS, Tsai CY, Huang HW, Chi HC, Lin YH et al (2016a) Potential diagnostic, prognostic and therapeutic targets of microRNAs in human gastric cancer. Int J Mol Sci 17 (6):945

Tsai MM, Wang CS, Tsai CY, Huang CG, Lee KF, Huang HW et al (2016b) Circulating microRNA-196a/b are novel biomarkers associated with metastatic gastric cancer. Eur J Cancer 64:137–148

Wang J, Song YX, Wang ZN (2015) Non-coding RNAs in gastric cancer. Gene 560(1):1–8

Wang J, Zhang H, Zhou X, Wang T, Zhang J, Zhu W et al (2018) Five serum-based miRNAs were identified as potential diagnostic biomarkers in gastric cardia adenocarcinoma. Cancer Biomark 23(2):193–203

Watabe H, Mitsushima T, Yamaji Y, Okamoto M, Wada R, Kokubo T et al (2005) Predicting the development of gastric cancer from combining Helicobacter pylori antibodies and serum pepsinogen status: a prospective endoscopic cohort study. Gut 54(6):764–768

Watanabe Y, Kim HS, Castoro RJ, Chung W, Estecio MR, Kondo K et al (2009) Sensitive and specific detection of early gastric cancer with DNA methylation analysis of gastric washes. Gastroenterology 136(7):2149–2158

Wayman J, Hayes N, Raimes SA, Griffin SM (2000) Prescription of proton pump inhibitors before endoscopy: a potential cause of missed diagnosis of early gastric cancers. Arch Fam Med 9 (4):385

Whiting JL, Hallissey MT, Fielding JWL, Dunn J (1998) Screening for gastric cancer by Helicobacter pylori serology: a retrospective study. Br J Surg 85(3):408–411

Wu J, Li G, Wang Z, Yao Y, Chen R, Pu X, Wang J (2015) Circulating MicroRNA-21 is a potential diagnostic biomarker in gastric cancer. Dis Markers 2015:435656

Yoon H, Kim N, Lee HS, Shin CM, Park YS, Lee DH et al (2011) Helicobacter pylori-negative gastric cancer in South Korea: incidence and clinicopathologic characteristics. Helicobacter 16 (5):382–388

Yuan CL, Li H, Zhu L, Liu Z, Zhou J, Shu Y (2016) Aberrant expression of long noncoding RNA PVT1 and its diagnostic and prognostic significance in patients with gastric cancer. Neoplasma 63(3):442–449

Zeng C, Rothfuss J, Zhang J, Chu W, Vangveravong S, Tu Z et al (2012) Sigma-2 ligands induce tumour cell death by multiple signalling pathways. Br J Cancer 106(4):693–701

Zhang K, Shi H, Xi H, Wu X, Cui J, Gao Y et al (2017) Genome-wide lncRNA microarray profiling identifies novel circulating lncRNAs for detection of gastric cancer. Theranostics 7(1):213

Zheng H, Takahashi H, Murai Y, Cui Z, Nomoto K, Niwa H et al (2006) Expressions of MMP-2, MMP-9 and VEGF are closely linked to growth, invasion, metastasis and angiogenesis of gastric carcinoma. Anticancer Res 26(5A):3579–3583

Zhou X, Yin C, Dang Y, Ye F, Zhang G (2015) Identification of the long non-coding RNA H19 in plasma as a novel biomarker for diagnosis of gastric cancer. Sci Rep 5:11516

Zhou X, Wen W, Zhu J, Huang Z, Zhang L, Zhang H et al (2017) A six-microRNA signature in plasma was identified as a potential biomarker in diagnosis of esophageal squamous cell carcinoma. Oncotarget 8(21):34468

Part II
Advances in Biosensors and Detection Technologies for Gastrointestinal Cancer

Chapter 9
Biosensors and its Applications for Early Detection of Gastrointestinal Cancer

Deepthi Nammi and Nageswara Rao Reddy Neelapu

Abstract Biosensors are used for detection of cancer and diseases; and are the potential contributors and great promising tools for the treatment of cancer due to its sensitivity, reliability, and low cost. Protein biomarkers, protein profiles, post-translation modifications, and gene expression changes are some of the important molecular notations that paved a new path for the development of biomarkers and biosensors. The different types of biosensors are affinity biosensor, amperometric biosensor, catalytic biosensor, DNA biosensor, electrochemical biosensor, graphene-based biosensor, mass change biosensor, metabolism biosensor, microbial biosensor, miRNA biosensor, optical biosensor, and many more. Biosensors work based on the recognition of elements, signal transduction, and its biological response. Biosensors can be used for early detection of cancer, cardiac disease, diabetes, and many infectious diseases. This chapter will discuss about the biomarkers used for cancer, trending biosensors, applications of biosensors, and the role of biosensors for early detection of gastrointestinal cancer.

Keywords Biomarkers · Biosensors · Cancer · Early detection of cancer · Gastrointestinal cancer

9.1 Introduction

Cancer is uncontrolled growth and division of cells due to cellular changes, and sometimes the visible growth of cells is called a tumor. The different types of cancers are breast cancer, brain cancer, cervical cancer, colon rectal cancer, gastric cancer, gastrointestinal cancer, lung cancer, nasopharyngeal cancer, ovarian cancer, prostate cancer, etc. Biomarkers play a potential role in the treatment of cancer right from

---

D. Nammi · N. R. R. Neelapu (✉)
Department of Biochemistry and Bioinformatics, GITAM Institute of Science, Gandhi Institute of Technology and Management (GITAM) (Deemed-to-be-University), Visakhapatnam, Andhra Pradesh, India

P. Veera Bramhachari, N. R. R. Neelapu (eds.), *Recent Advancements in Biomarkers and Early Detection of Gastrointestinal Cancers*, Diagnostics and Therapeutic Advances in GI Malignancies, https://doi.org/10.1007/978-981-15-4431-6_9

screening, detection, staging, prognosis, and diagnosis. In 1965, Dr. Joseph Gold established a test for a common type of cancers (Gold and Freedman 1965a). In 1970, serum tests were developed for a variety of cancers (Gold and Freedman 1965b). In 1980, biomarkers were developed for ovarian cancer, breast cancer, pancreatic cancer, colorectal cancer (CA-125, CA15-3, CA19-9), and prostate cancer (prostate-specific antigen) (Yilmaz et al. 2001). This motivated researchers to identify suitable biomarkers for different types of cancers (Chatterjee and Zetter 2005). This chapter will discuss and provide information on biomarkers for cancer, biosensors and its applications, the role of biomarkers, and biosensors in early detection of gastrointestinal cancer.

9.2 Biomarkers and Cancer

A molecule secreted by a tumor or specific response due to the presence of cancer is referred to as biomarkers. These biomolecules are present in serum, plasma, and tissues that are useful for therapeutic interventions against varieties of cancers like breast cancer, brain cancer, cervical cancer, colon rectal cancer, gastric cancer, gastrointestinal cancer, lung cancer, nasopharyngeal cancer, ovarian cancer, and prostate cancer, etc. Biomarkers can act as indicators to measure and evaluate the normal biological process or pathogenic process or biological response to therapeutic intervention in cancer. Biomarker helps in early detection of disease and its risk, prediction of disease recurrence, and monitoring treatment of drugs (Ray et al. 2011). Therefore, the identification of biomarkers is worthy for the early detection of aggressive diseases like cancer. Biomarkers are classified based on disease state, types of biomolecules, and other criteria (Fig. 9.1). The disease state biomarkers include prediction biomarkers, detection biomarkers, diagnostic biomarkers, and prognosis biomarkers (Fig. 9.1). The biomolecule biomarkers include DNA biomarker, RNA biomarker, protein biomarker, and glycol biomarkers (Fig. 9.1). The biomarkers in other criteria include imaging biomarkers, pathological biomarker, and in silico biomarkers (Fig. 9.1).

9.2.1 Biomarkers for Lung Cancer

Lung cancer is the second most common cancer in the world. The malignant tumor that may arise from different cell types of the lung with histological variants is referred to as lung cancer (Zamay et al. 2017). Lung cancer is mainly of two main types—small cell lung carcinoma and non-small cell lung carcinoma. Approximately 80% of lung cancers are non-small cell lung carcinomas. According to the American cancer society, 228,150 new cases were recorded in both men and women for the year 2019. About 142,670 death cases were recorded in both men and women due to lung cancer (Cancer Facts and Figures 2019). Early detection of lung cancer is

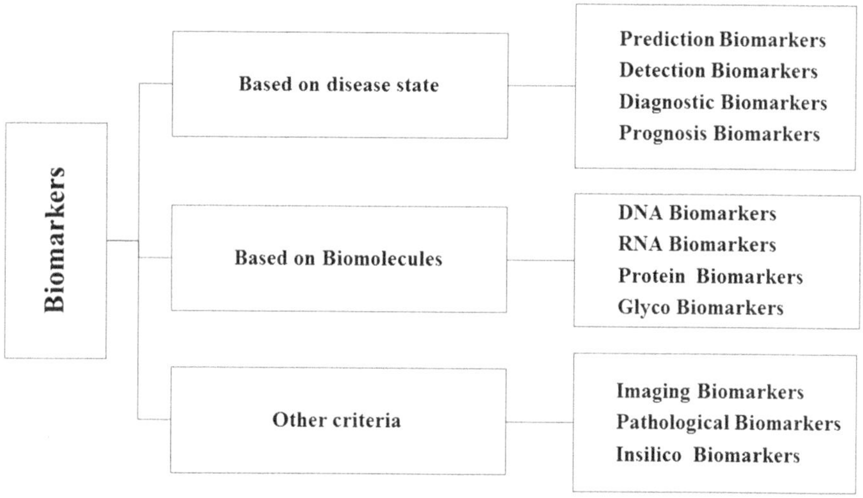

Fig. 9.1 Classification of biomarkers

a very effective and highly challenging; and identifying novel, potential, highly sensitive and specific biomarkers are needed (Zamay et al. 2017). Biomarkers identified to date for lung cancer are listed in Table 9.1. Serum amyloid A, apolipoprotein A 1, and KLKB1 are the potential lung cancer biomarkers. The normal level of serum amyloid A in a healthy individual is 2 μg/mL, whereas in diseased individuals it is elevated to 62.4 μg/mL. Apolipoprotein A-1 fragment and 17–18 kDa fragment of plasma kallikrein B1 is down-regulated in diseased states when compared to healthy individuals (Sung and Cho 2008; Cho et al. 2004; Maciel et al. 2005). Thus, these specific biomarkers can be employed for the diagnosis of lung cancer.

9.2.2 Biomarkers for Ovarian Cancer

The most common cancer in women is ovarian cancer. The female reproductive system contains two ovaries, and these small organs are connected to the womb. Women in menopause stage are prone to ovarian cancer (Berge et al. 2018). According to the American cancer society, 22,530 new cases were estimated in the year 2019. About 13,980 death cases were estimated due to ovarian cancer (Cancer Facts and Figures 2019). The detection of ovarian cancer is very difficult until it spreads to the pelvis and abdomen. At the same time, it is very difficult to treat ovarian cancer at the last stage. The different types of ovarian cancer are—epithelial tumors, germ cell tumors, and stromal tumors (Berge et al. 2018). Biomarkers identified to date for ovarian cancer are listed in Table 9.1. Cancer antigen 125 (CA125) is well characterized and established biomarker for ovarian cancer

Table 9.1 Potential biomarkers for cancer

S. no.	Types of cancers	Biomarkers	References
1.	Ovarian cancer	CA125	Rai et al. (Berge et al. 2018)
2.	Ovarian cancer	Apolipoprotein A1, transthyretin, inter-trypsin inhibitor heavy chain H4	Zhang et al. (2004)
3.	Ovarian cancer	Osteopontin	Kim et al. (2002)
4.	Ovarian cancer	KLK 7, 10, 11, 13	Zheng et al. (2007)
5.	Ovarian cancer	B7-H4	Shih et al. (2007)
6.	Tumor	Survivin	Duffy et al. (2007)
7.	Prostate cancer	Matriptase	Saleem et al. (2006)
8.	Prostate cancer	KLK2	Magklara et al. (1999)
9.	Nasopharyngeal cancer	Serum amyloid A	Cho et al. (2004)
10.	Lung cancer	Haptoglobin-α2	Maciel et al. (2005)
11.	Lung cancer	APOA1	Maciel et al. (2005)
12.	Lung cancer	KLKB1	Heo et al. (2007)
13.	Lung cancer	Annexin	Zamay et al. (2017)
14.	Lung cancer	Vimentin	Zamay et al. (2017)
15.	Lung cancer	Thymosin	Zamay et al. (2017)
16.	Lung cancer	Cofilin	Zamay et al. (2017)
17.	Lung cancer	Serum ameloid	Sung and Cho (2008)
18.	Lung cancer	MBL2	Zamay et al. (2017)
19.	Lung cancer	AAG1-2	Zamay et al. (2017)
20.	Lung cancer	FGA	Zamay et al. (2017)
21.	Lung cancer	FIBA	Zamay et al. (2017)
22.	Lung cancer	CAV1	Zamay et al. (2017)
23.	Lung cancer	GSN	Goetsch (2011)

(continued)

Table 9.1 (continued)

S. no.	Types of cancers	Biomarkers	References
24.	Lung cancer	FCN3	Goetsch (2011)
25.	Lung cancer	CNDP1	Goetsch (2011)
26.	Lung cancer	UCRP	Foa et al. (1999)
27.	Lung cancer	ALB	Foa et al. (1999)
28.	Lung cancer	IGFBP7	Foa et al. (1999)
29.	Lung cancer	MMP14	Foa et al. (1999)
30.	Lung cancer	THBS1	Foa et al. (1999)
31.	Non-small cell lung cancer	K-ras, P53	Aviel-Ronen et al. (2006)
32.	Lung cancer	K-ras, P53, serine protease family member-trypsinogen IV (PRSS3), tissue inhibitor of metalloproteinase (TIMP)-3, death-associated protein (DAP)-kinase P16, FHIT	Aviel-Ronen et al. (2006)
33.	Lung cancer	Haptoglobin-α2	Aviel-Ronen et al. (2006)
34.	Lung cancer	APOA1	Aviel-Ronen et al. (2006)
35.	Lung cancer	KLKB1	Aviel-Ronen et al. (2006)
36.	Lung cancer	Dihydrodiol dehydrogenase	Li et al. (2015a)
37.	Lung cancer	KLK 4, 8, 10, 11, 12, 13, 14	Planque et al. (2008)
38.	Cancer	5-Hydroxymethylcytosine	Fukushige and Horii (2013)
39.	Cancer	Auto antibody	Nesterova et al. (2006)
40.	Colon rectal cancer	Multigene profiles	Salazar et al. (2011)
41.	Colon rectal cancer	TIMP-1	Birgisson et al. (2010)
42.	Colon rectal cancer	CA19-9	Yakabe et al. (2010)
43.	Colon rectal cancer	*SEPT9*	Warren et al. (2011)
44.	Colon rectal cancer	Laminin	Wu et al. (2008)
45.	Colon rectal cancer	Collapsin response mediator	Wu et al. (2008)
46.	Colon rectal cancer	Prostatic acid phosphatase	Wu et al. (2008)
47.	Colon rectal cancer	Protein-2 (CRMP-2)	Wu et al. (2008)

(Rai et al. 2002). There is an increase (nearly 80%) of CA-125 levels in epithelial ovarian cancer patients when compared with healthy individuals. There is also a difference between pregnant and non-pregnant women's CA-125 levels. An increase of 16% CA-125 levels is observed in the first trimester of pregnant women and also a high concentration of CA-125 is seen in amniotic fluid. High levels (greater than 65 U/mL) of CA-125 were observed in non-pregnant women with gynecological disorders (Niloff et al. 1984). Thus, biomarker CA-125 can be employed in the early detection of ovarian cancer.

9.2.3 Biomarkers for Prostate Cancer

The prostate is a small gland, situated below the bladder near the rectum in the male reproductive system, and it enriches the sperm by producing fluid that makes semen. Prostate cancer is leading cancer that causes death in men. In 2019, 74,650 cases were estimated among them 31,620 deaths are associated with prostate cancer (Cancer Facts and Figures 2019). The patterns or changes in the metabolic biofluids help in identifying biomarkers for prostate cancer (Gómez-Cebrián et al. 2019). The broad categories of biomarkers for prostate aggressive cancer include blood-based biomarkers, serum-based biomarkers, tissue-based biomarkers, and urine-based biomarkers. Biomarkers identified to date for prostate cancer are listed in Table 9.1. Some of the potential biomarkers are acetyl histidine, citrulline, choline, glycerol-3-phosphate, prostate-specific antigen (PSA), and tyrosine (Kdadra et al. 2019). The United States Preventative Services Task Force (USPSTF) in the USA during 2012, recommended prostate-specific antigen (PSA) as a potential tool for screening prostate cancer (Rice and Stoyanova 2018). The normal level of PSA is 0–4 ng/mL, whereas the elevated levels of PSA greater than 0-4 ng/mL are considered suspicious of prostate cancer. PSA levels vary with age and therefore, elevated levels of PSA greater than 0–7.5 ng/mL should be considered for prostate cancer (Rice and Stoyanova 2018). Thus, PSA can be considered as a good biomarker for the early detection of prostate cancer.

9.2.4 Biomarkers for Colon Rectal Cancer

Colon rectal cancer or bowel cancer is the third most common cancer death in both men and women. The statistics of colon rectal cancer shows that the worldwide burden of colon rectal cancer has increased by 60%. The American Cancer Society (ACS) estimates 101,420 new cases for colon cancer and 44,180 rectal cancer cases in the year 2019 (Cancer Facts and Figures 2019). According to the Irish Cancer

Society, colon rectal cancer is one of the topmost cancers and the estimated incidence of 2016–2018 in 1136 cases in women and 1631 cases in men (Cancer in Ireland 1994–2016 with estimates for 2016–2018 n.d.). Colon rectal cancer starts in the inner lining of colon or rectum (called colon rectum) as an abnormal growth of the inner lining of tissue. This abnormal growth is called polyps and some polyps are cancerous. Polyps are of two types adenomatous polyps and hyperplastic polyps or inflammatory polyps. Mutations in genes, family history, and obesity, etc. are the causes of colon rectal cancer. Extensive studies were carried out on colon rectal cancer to identify potential biomarkers for early detection and better treatment. Biomarkers identified to date for colon rectal cancer are listed in Table 9.1.

9.2.5 Biomarkers for Gastrointestinal Cancer

Gastrointestinal cancer or sometimes known as adenocarcinoma develops in the mucous lining of the stomach. Cancer in different sections shows different symptoms and need different treatments. Gastrointestinal cancer is the most leading cancer of death in the world. Identification of biomarkers for gastrointestinal cancer helps in early diagnosis and efficient monitoring of the disease. Gastrointestinal cancer biomarkers are broadly classified as noninvasive biomarkers such as blood biomarkers and gastric juice-based biomarkers (Matsuoka and Yashiro 2018). Bio markers identified to date for gastrointestinal cancer are listed in Table 9.2. The serum miRNA is a potential biomarker and three major strong serum potential biomarkers are miR221, miR376C, and miR374 for early detection of gastrointestinal cancer (Song et al. 2012). HER2 is overexpressed in gastrointestinal cancer and thus can be used as a potential biomarker for early detection of gastrointestinal cancer. KIT gene is universally expressed, mutations in the exon region of the KIT gene leads to gastrointestinal tumors. The mutated KIT gene can be used as a biomarker for gastrointestinal tumors. This gene is present in abdominal tumors and non-abdominal tumors (breast cancer and melanomas) (Duffy et al. 2014). Carcinoembryonic antigen (CEA) is the potential biomarkers for gastrointestinal cancer. The normal range of gastrointestinal cancer biomarker CEA is between 0 and 3 ng/mL (Asao et al. 1991). The levels of CEA in stage I, stage II, stage III, and stage IV is 16.3, 48.91, 57.99, and 66.8, respectively (Elimova et al. 2015). The levels of CA are elevated, i.e., >100 ng/g in gastric cancer patients when compared with healthy individuals. Thus, potential biomarkers like miR221, miR376C, miR374, HER2, KIT gene, and CEA can be employed for early detection of gastrointestinal cancer.

Table 9.2 Potential biomarkers in gastrointestinal cancer

S. no.	Biomarkers	References
1.	HER2 (ERBB2)	Elimova et al. (2015)
2.	EGFR	Elimova et al. (2015)
3.	VEGFA	Elimova et al. (2015)
4.	NOTCH1	Elimova et al. (2015)
5.	p-mTOR	Elimova et al. (2015)
6.	MMP1, MMP7	Elimova et al. (2015)
7.	TGFB1	Elimova et al. (2015)
8.	MET	Elimova et al. (2015)
9.	HER3 (ERBB3)	Elimova et al. (2015)
10.	SHH/PTCH1/SMO	Elimova et al. (2015)
11.	FGFR2	Elimova et al. (2015)
12.	CASOX9	Elimova et al. (2015)
13.	TP53	Elimova et al. (2015)
14.	PTEN	Elimova et al. (2015)
15.	ALDH	Elimova et al. (2015)
16.	PIK3	Elimova et al. (2015)
17.	PD-L1	Curea et al. (2017)
18.	ADAM23	Watanabe et al. (2009)
19.	GDNF	Watanabe et al. (2009)
20.	MINT25	Watanabe et al. (2009)
21.	MLF1	Watanabe et al. (2009)
22.	PRDM5	Watanabe et al. (2009)
23.	RORA	Watanabe et al. (2009)
24.	BARHL2	Yamamoto et al. (2016)
25.	PVT1	Yuan et al. (2016)
26.	CagA	Saju et al. (2016)
27.	VacA	Ghotaslou et al. (2018)
28.	Gastrokine 1	Altieri et al. (2017)
29.	CEACEM6	Yasui et al. (2004)
30.	APOC1	Yasui et al. (2004)
31.	YF13H12	Yasui et al. (2004)
32.	CDH17	Yasui et al. (2004)
33.	OLFM4	Oue et al. (2015)
34.	HOXA10	Oue et al. (2015)
35.	DSC2	Oue et al. (2015)
36.	TSPAN8	Oue et al. (2015)
37.	TM9SF3	Oue et al. (2015)
38.	FUS	Yasui et al. (2004)
39.	COLIA1, COLIA2	Yasui et al. (2004)
40.	APOE	Yasui et al. (2004)
41.	CCNB1 and CCNB2	Wang et al. (2015)
42.	ZNF331	Marie Vedeld et al. (2015)

(continued)

Table 9.2 (continued)

S. no.	Biomarkers	References
43.	ZSCAN18	Marie Vedeld et al. (2015)
44.	CDO1	Marie Vedeld et al. (2015)
45.	KLK6	Paliouras et al. (2007)
46.	CD44v6, MM-7	Okayama et al. (2009)
47.	VEGF	Maeda et al. (1995)
48.	Long non-coding RNA and MicroRNA	Afsane et al. (Bahrami et al. 2018)
49.	miR-21	Chan et al. (2008)
50.	Carbohydrate antigen 19-9	Duffy (1998)
51.	hsa_circ_002059	Li et al. (2015b)
52.	REGIV	Yasui et al. (2004)

9.3 Biosensors

The conversion of biological response into an electrical signal is referred to as "Biosensor." The important and salient features of biosensors are its cheap, portable, capable, purposeful, accurate in response, reproducible, electrical disturbance-free, non-fouling, non-proteolysis, highly specific, stable under normal conditions, etc. Biosensors have two important components, i.e., biological component and physical component. Enzyme, metabolites, etc. are the biological components, whereas amplifier and transducer are the physical components (Malhotra et al. 2017). The biological component interacts with the analyte to produce a physical change, i.e., signal. The analyte is a compound or a substance whose chemical constituents are being identified and measured. This response is detected by the transducer, and the response is amplified by an amplifier, followed by processing via a processor and then displayed by a displayer (Malhotra et al. 2017). There are three generations of biosensors—first-generation biosensors, second-generation biosensors, and third-generation biosensors. The first-generation biosensor detects the response caused by the normal product of the reaction and diffuses to the transducers. The second-generation biosensor uses specific mediators between reactions and transducers. The third-generation biosensor uses no mediator and the reaction itself causes a response. Figure 9.2 demonstrates the developments and the role of biosensors in health care monitoring.

9.3.1 The Principle and Working of Biosensor

The desired biological material like an antibody, enzymes, whole-cell, hormones, nucleic acids, etc. are immobilized using conventional methods like covalent and non-covalent binding. The sample (biological material) is passed through the membrane and some are retained as intrusive molecules outside the membrane. The biological sample interacts with the biosensor and forms a product. The product

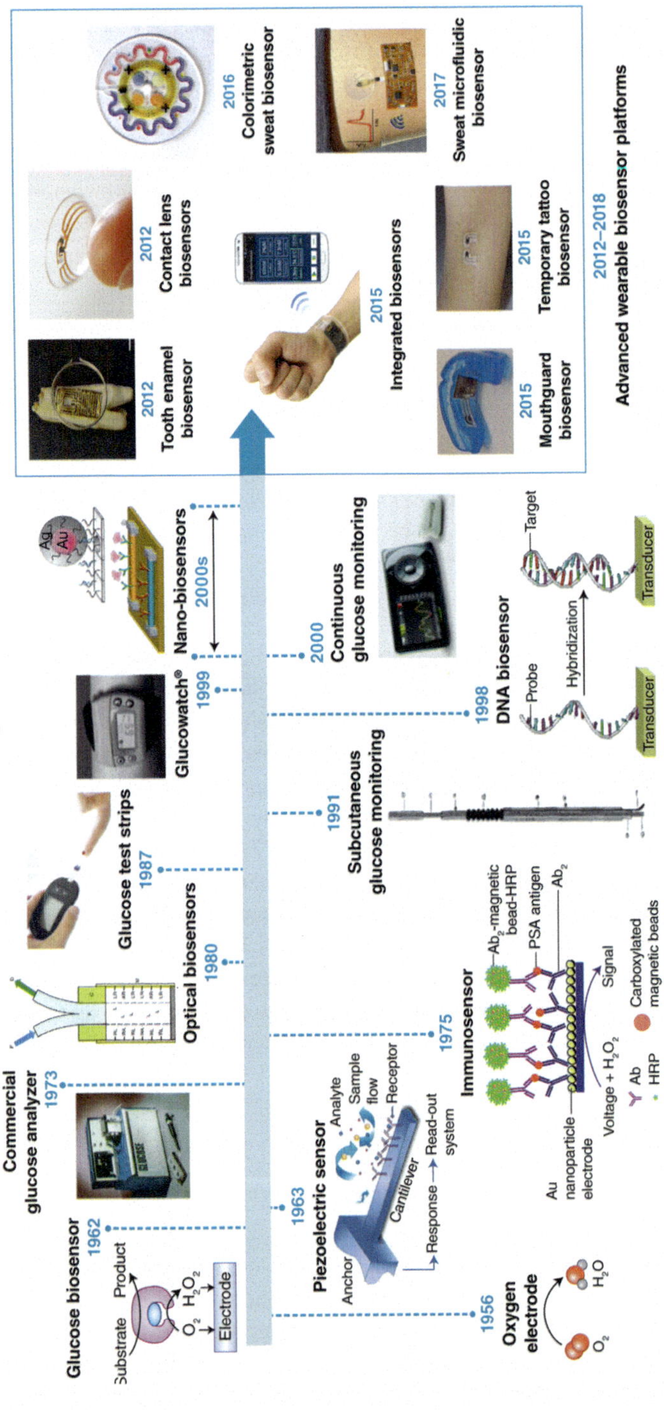

Fig. 9.2 Developments of wearable biosensors for health care monitoring (Source: Kim et al. (2019))

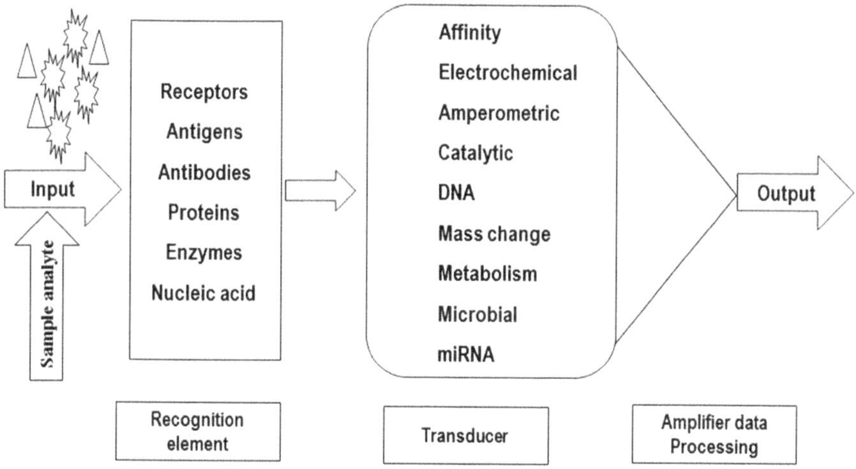

Fig. 9.3 Working principle of biosensors demonstrating flow from the sample, input, recognition elements, transducers, processing of data, and output

may be a charge, chemical, current, heat, and gas (Malhotra et al. 2017). The product passes through another membrane and reaches the transducer. Along with the transducer, amplifier, recorder, and display devices are the required components of biosensors (Fig. 9.3).

9.3.2 *Important Attributes of Biosensors*

The important attributes of biosensors are selectivity, stability, sensitivity, linearity, and reproducibility. Selectivity is one of the most important characteristics of biosensor and biosensor detects the analyte specifically in the sample. The antigen–antibody interaction is one of a good example of selectivity, where selectivity is the major step in selecting antibodies for biosensors and the immobilized antibody interacts with antigen in solution (Bhalla et al. 2016). Stability is an important feature for sensors in the sensing system which is required for continuous monitoring. Transducer response, temperature, affinity of the receptor, the extent of analyte binding to the receptor, and sensitivity influence stability. The high binding affinity shows a strong electrostatic bonding of the receptor and analyte that fortifies biosensor stability (Bhalla et al. 2016). Sensitivity is an essential attribute for biosensor and is also known for its threshold and limit of detection. The amount of analyte in the sample can be detected by a biosensor based on its sensitivity. Therefore, biosensors can detect biomarkers in the sample based on sensitivity (Bhalla et al. 2016). Linearity is one of the important attributes of a biosensor which measures the response accurately with a mathematical representation $y = mc$; where m is sensitivity, c is the concentration of the analyte, and y is signal output. The concentration of the analyte is associated with biosensor resolution, which is based on the changes in the response of biosensor resolution, along with the concentration (Bhalla et al. 2016). Reproducibility is one of the major attributes and the ability of biosensor to generate similar responses when an experiment is repeated is essential. The accuracy and precision of the transducer increase the ability of biosensors to produce similar responses every time. This indicates that the mean value is close to true value when repeatedly measured. Robustness and reliability of the biosensor are provided by reproducibility (Bhalla et al. 2016).

9.4 Types of Biosensors

Literature reports different types of biosensors like affinity biosensors, amperometric biosensors, catalytic biosensors, DNA biosensors, electrochemical biosensors, graphene-based biosensor, mass change biosensor, metabolism biosensor, microbial biosensor, miRNA biosensor, optical biosensor, and many more.

9.4.1 Affinity Biosensors

Molecules like antibodies, nucleic acids, and hormones bind to analyte irreversibly causing physical changes. These changes lead to the formation of complex or disassociation and are detected by the transducer. Some parameters for affinity biosensors are sensitivity, specificity kinetic parameters of affinity interactions, stability, sensor material, type of transducer, affinity elements, etc. (Leech 1994). To detect the antigens in the sample, affinity sensors use two different assays—direct affinity assay or competitive assay.

9.4.1.1 Direct Affinity Assay

The antigen (analyte) in the sample is detected using reporter, recognition element, and enzyme to the label. Direct affinity assay or sandwich assay applies to large molecular weight analytes, and it is not useful for small molecular weight analytes. The analyte concentration is directly proportional to signal, and the interpretation of direct assay is based on signal intensity. If the signal intensity is high, then the sample is with high analyte concentration, whereas if the intensity of the signal is low, then the sample is with no analyte (Leech 1994).

9.4.1.2 Competitive Assay

The competitive assay helps in detecting the presence of the analyte and binding sites in the sample. The signal is reversely proportional to analyte concentration, and this assay is suitable for low molecular weight analytes. The interpretation of competitive affinity assay depends on signal intensity. If the signal intensity is high, then there is no analyte in the sample, whereas low signal intensity is seen in samples with a high concentration of analyte (Leech 1994). The association and disassociation constant describes the binding affinity of the assay.

9.4.2 Electrochemical Biosensor

The most common type of biosensor is electrochemical biosensor, due to its size (small), effectiveness, and user-friendly nature. The electrochemical biosensors are based on enzyme catalysis reactions it produces or consumes electrons (Figs. 9.4 and 9.5). The sensors contain three electrodes—a working electrode, a counter electrode, and a reference electrode. In 1962, Leland C. Clark introduced the first enzyme electrode with immobilized glucose oxidase. The sensing molecules are either coated or bonded to a probe surface and with the membrane in a place, excluding analyte solution. These sensing molecules react with compounds and are detected by the electrical signal which is based on the concentration of the analyte. The electrochemical biosensor converts the chemical information to a measurable amperometric signal by the amperometric transducer or potentiometric transducer or impedimetric transducer. The distinct and the most commonly used techniques for the detection of biomarkers are electrochemical biosensor are cyclic voltametry, electrochemical impedance spectroscopy, linear sweep voltametry, square wave voltametry, and stripping voltametry. There are two different types of electrochemical biosensors—amperometric biosensor and potentiometric biosensor. The electrical response of a specific element and molecule is recognized or detected by the potentiometric

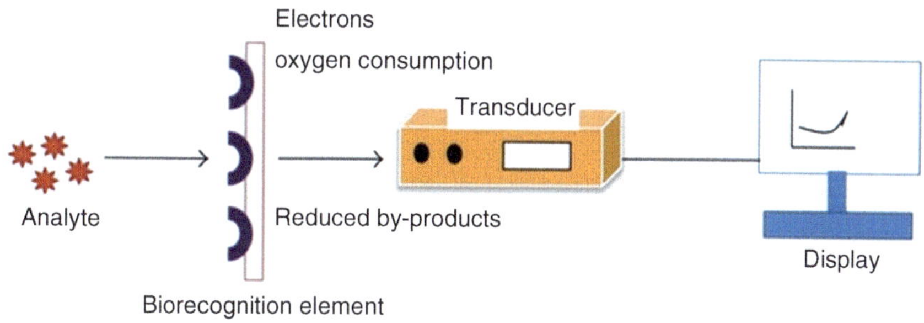

Fig. 9.4 Prototype model of an electrochemical biosensor which detects the analyte in the sample by recognizing electrons consumed or produced and transduce this signal to the displaying unit (Source: Dhull et al. (2013))

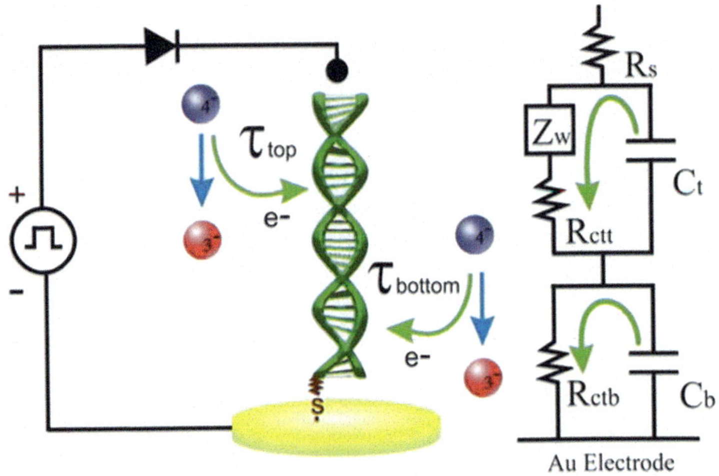

Fig. 9.5 Prototype model of DNA electrochemical biosensor (Source: Espinosa et al. (2019))

biosensor. The electrodes produce a signal and the signal produces oxidoreduction reaction which is measured by amperometric biosensors. The amperometric biosensors use DNA sequences for cancer detection, and these sensors are very potential in prognosis and diagnosis of cancer (Freitas et al. 2018).

9.4.3 Optical Biosensors

The variations in the wavelength can be measured by optical biosensors (Fig. 9.6). The optical transducers having photonic crystals convert the change in the wavelength in response to the cognition of the analyte. The light areas and small volumes are captured and the results are transmitted to a high magnetic field, where the association and disassociation of molecules to crystal surface are measured using biosensors. The optical biosensor can detect or monitor the changes in proliferation, apoptosis, and their role in cancer (Jainish and Prittesh 2017).

9.4.4 Calorimetric Biosensor

The exothermic reactions in cancer and normal cells are detected or measured using a calorimetric biosensor (Fig. 9.7). The changes in temperatures of the desired molecules during the enthalpic reaction are measured (Medley et al. 2008). The differences in the temperatures of both cells can be measured by calorimetric biosensor and help detect cancer cells and normal cells.

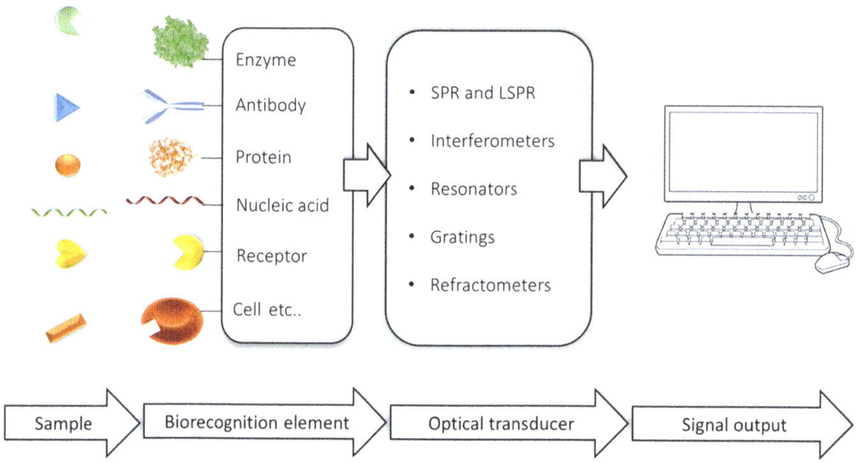

Fig. 9.6 Prototype model for optical biosensors (Source: Damborský et al. (2016))

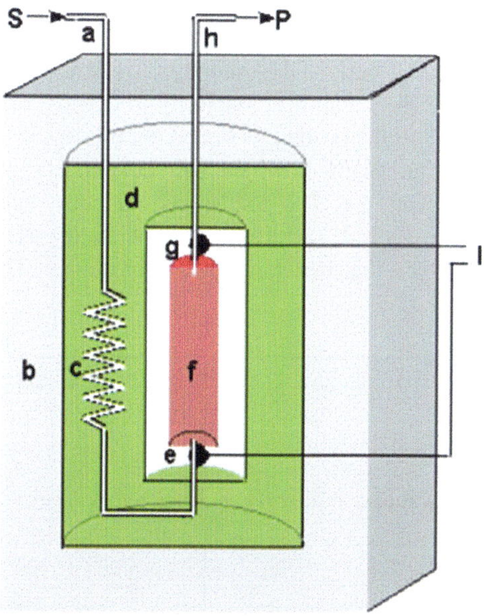

Fig. 9.7 Schematic diagram of a colorimetric biosensor. The sample stream passes (**a**) through the outer insulated box (**b**) to the heat exchanger (**c**) within an aluminum block (**d**). From there, it flows past the reference thermistor (**e**) and into the packed bed bioreactor (**f**, 1 mL volume), which contains the biocatalyst, where the reaction occurs. The change in temperature is determined by the thermistor (**g**) and the solution passed to waste (**h**). External electronics (**l**) determine the difference in the resistance, and hence temperature, between the thermistors (Source: Tiquia-Arashiro (2014))

9.4.5 Mass-Sensitive Biosensors

The mass-sensitive biosensors are of two types—acoustic biosensor and piezoelectric biosensor. These biosensors rely on the changes of quartz crystal mass when potential energy is applied to the quartz crystal. The changes in mass are converted to signal and are detected by mass-sensitive biosensors. Mass-sensitive biosensors help detect different types of tumor biomarkers (Tothill 2009).

9.4.6 DNA Biosensor

Oligonucleotide sequences or DNA/RNA fragments based on nucleic acids are widely used as biosensors. DNA biosensors are based on the hybridization of specific DNA/RNA strands. DNA biosensors are faster in detection, cheaper in terms of cost, simpler, specific, and can be used multiple times (Fig. 9.8). DNA biosensors are very sensitive when it is combined with a polymerase chain reaction. DNA hybridization biosensors are based on the complementary base pairs of short sequences that are used as selective DNA segments that are immobilized to electrode surface to retain accessibility, stability, reactivity to optimal orientation, and target analyte. The target binds to the probe DNA, and the process is called hybridization. Hybridization is measured with enzymes like horseradish, peroxidase, alkaline phosphatase, colloidal gold, etc. The conditions like time, temperature, and ionic strength are important to achieve sensitivity and specificity (Kavita 2017). Peptide-nucleic acid has opened another new era of DNA biosensor. The pseudo peptide is

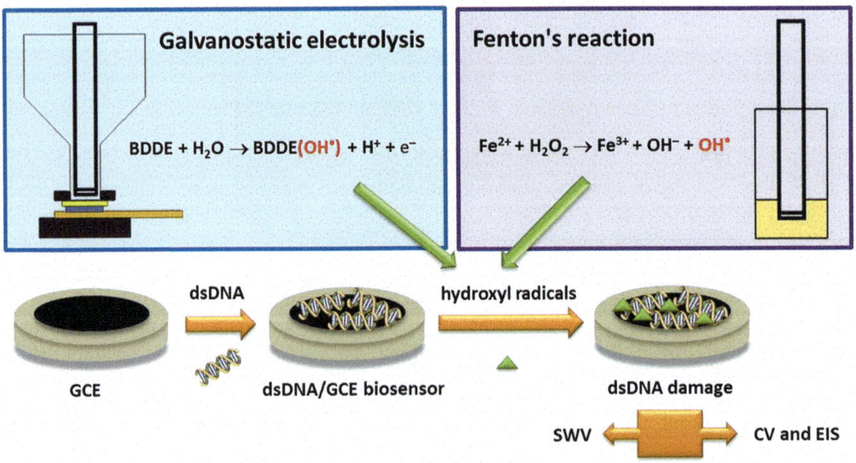

Fig. 9.8 Electrochemical DNA biosensor for the detection of DNA damage induced by hydroxyl radicals (Source: Hájková et al. (2017))

placed in sugar-phosphate in connection with DNA biosensors which helps to detect the single-base mismatches.

9.4.7 *Microbial Biosensor*

Microbial biosensor uses enzymes of microorganisms as a recognition element, and these enzymes can produce a highly selective and specific response with the analyte. This device transduces and detects the target analytes when microorganisms are immobilized. This technology is more reliable and selective and most helpful for the analysis of clinical and environmental samples. The different types of microbial biosensors are optical biosensors with optical transducers that produce changes in optical properties like fluorescence, absorption, luminescence, and refractive index (Lei et al. 2006). The different types of microbial biosensors are fluorescent microbial biosensor, bioluminescence microbial sensor, colorimetric microbial biosensor, and microbial fuel cells (MFCs) biosensor.

9.4.7.1 Fluorescent Microbial Biosensor

The fluorescent microbial biosensor is widely used for the detection of toxic compounds, where the fluorescent light is directly proportional to the concentration of analytes at a low level (D'Souza 2001). The green fluorescent protein is most commonly used in microbial biosensor protein which encodes a promoter to a reporter gene that emits fluorescence in genetically engineered microorganisms (Su et al. 2011). This fluorescent microbial biosensor is used for the detection of toxic compounds (García-Alonso et al. 2009).

9.4.7.2 Bioluminescence Microbial Biosensor

Bioluminescence microbial sensor is used for the detection of toxicity in the environment (Steinberg et al. 1995). Bioluminescence microbial biosensor measures the density changes of the bioluminescence in the living cells in proportion to the concentration of analytes. According to the method, the production of bioluminescence is the expression of the lux gene in two ways, one is the inducible mode and the other is a constitutive manner.

9.4.7.3 Colorimetric Microbial Biosensor

The colorimetric microbial biosensor detects analytes and their concentration in the sample when the compound changes its color. For example, *Flavobacterium* sp. was constructed as a colorimetric microbial biosensor with a microbial transducer, where the bacterium hydrolyzes and forms a chromophoric product known as methyl parathion. The microbial transducer detects methyl parathion (Kumar et al. 2006).

9.4.7.4 Microbial Fuel Cells (MFCs) Biosensor

MFC biosensor via microbial catabolism converts organic substrates into electricity. MFCs biosensor contains two chambers: a cathodic and anodic chamber for proton exchange. The microbe oxidizes the fuel in the anodic chamber, where the generated electrons and protons are transferred to the cathodic chamber by electric circuit externally. The biosensors are used to measure water toxicity and biochemical oxygen demand due to its fast response and stability (Choi and Chae 2012).

9.4.8 MicroRNA Biosensor

A class of small non-coding RNA molecules that regulate a biological process by regulating gene expression are known as microRNAs. Many of these molecules have a key role in controlling infection concerning immunity. Differential expression of genes is a valuable source for the identification of novel biomarkers for infectious diseases and cancers (Correia et al. 2017). Detection and amplification of free microRNA are based on different modes such as amplification-based polymerase methods, optical methods, and electrochemical methods of biosensors. The advantages of miRNA-based biosensors are "Limit Of Detection (LOT)," robustness and "Time To Results (TTR)."

9.4.9 Metabolite Biosensors

Metabolite sensors gained importance as they can detect enzymes or molecules that are produced inside the cell (Morgan et al. 2016; Rogers et al. 2016; Liu et al. 2017; Cheng et al. 2015; Kwon et al. 2018). Metabolite sensors are made of modules like sensing module and reporter module. The sensing module is made of a transcriptional regulator to detect a ligand. The reporter module consists of a reporter gene that reports the transcription of the gene, and this output is measured as a signal (Chong and Ching 2016). Frazao et al. (2018) designed and implemented a metabolite sensor for the detection of aldehydes coupled with FACS-based selection.

9.4.10 *Forster Resonance Energy Transfer (FRET) Biosensors*

FRET-based biosensors consist of a pair of fluorophores—acceptors and donors (Zhang et al. 2014a). The domain-binding ligand is placed in between two fluorophores, and the conformational changes occur when a ligand binds to target leading to a change in the FRET signal (Bermejo et al. 2011). The ligand-binding proteins could be regulatory proteins, periplasmic proteins, and ligand sensing proteins (Peroza et al. 2015; Mohsin et al. 2013). This biosensor helps to detect a wide variety of small molecules.

9.4.11 *Graphene-Based Biosensor*

Graphene-based biosensors are made up of graphene materials and are superior to nanomaterials, and these are commercially limited. Excellent conductivity, high surface to volume ratio, and small gap band for reading outs of electrochemical and optical readouts are the advantages with graphene-based biosensors (Szunerits and Boukherroub 2018). The graphene biosensors are prepared from different derivatives of grapheme like Graphene Oxide (GO), Reduced Graphene Oxide (rGO) and partially reduced Graphene Oxide (prGO) nanosheets. The methods used for the preparation of grapheme derivatives are chemical vapor deposition methods (Zagorodko et al. 2014); synthetic approach by the weakening of van der Waals forces between grapheme layers; spin-coating, drop-casting, interaction between positively and negatively charged nanosheets; electrochemical reduction; and electrophoretic deposition (Sun et al. 2013). The speed, sensitivity, and selectivity made graphene biosensor ideal for the development of a medical diagnostic test for the Zika virus (Canbaz and Sezgintürk 2014). Afsahi et al. (2018) developed graphene-based biosensor, immobilized with the Zika virus antibody, to detect native Zika viral (ZIKV) antigens. The biosensor was able to detect antigen ZIKV NS1 in the sample at 450 pM.

9.5 Applications of Biosensors

Biosensors play a vital role in the detection of disease, identification of drug targets, the discovery of drugs, monitoring of the environment, and ensuring the safety of food, detection of toxic chemical and biological agents of defense interest, etc. to improve the quality of life, stability, and sensitivity (Arora et al. 2011). Some biosensors work as single-shot analysis tools that are cost-effective and other biosensors functions as a long-term analysis tools where it takes hours to several days

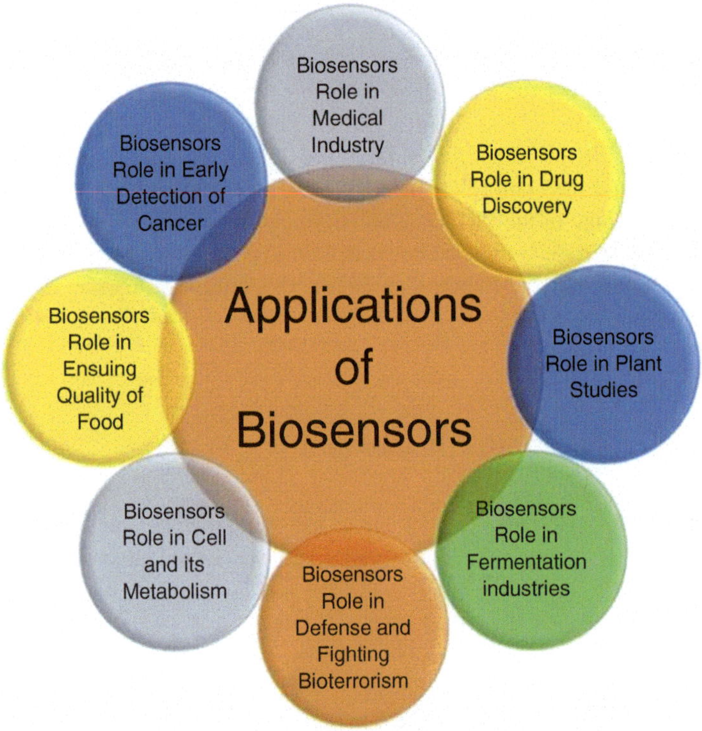

Fig. 9.9 Applications of biosensors in different facets like detection of disease, identification of drug targets, the discovery of drugs, monitoring of the environment, safety of food, detection of toxic chemical and biological agents of defense interest

(Van Dorst et al. 2010). The role of biosensors in different areas is discussed below (Fig. 9.9).

9.5.1 *Biosensors Role in Medical Industry*

Biosensors have an important role in the medical industry, mostly in diagnosing diseases. Biosensors' role in the detection of glucose, diagnosing critical infectious disease, identification of pathogens, detecting antimicrobial activity, identifying end-stage of heart failure in patients, adverse effects of early phase ventricular problems, detection of cytokines, detection of antigen-antibody interactions, and early detection of cancer is established. A few examples will be discussed to understand the role of biosensors in the medical industry (Fig. 9.10). The glucose sensor is a major outbreak in the field of clinical medicine, which is used in diagnosing diabetes mellitus to ensure the control of blood glucose levels (Scognamiglio et al. 2010). The household usage of glucose biosensors accounts

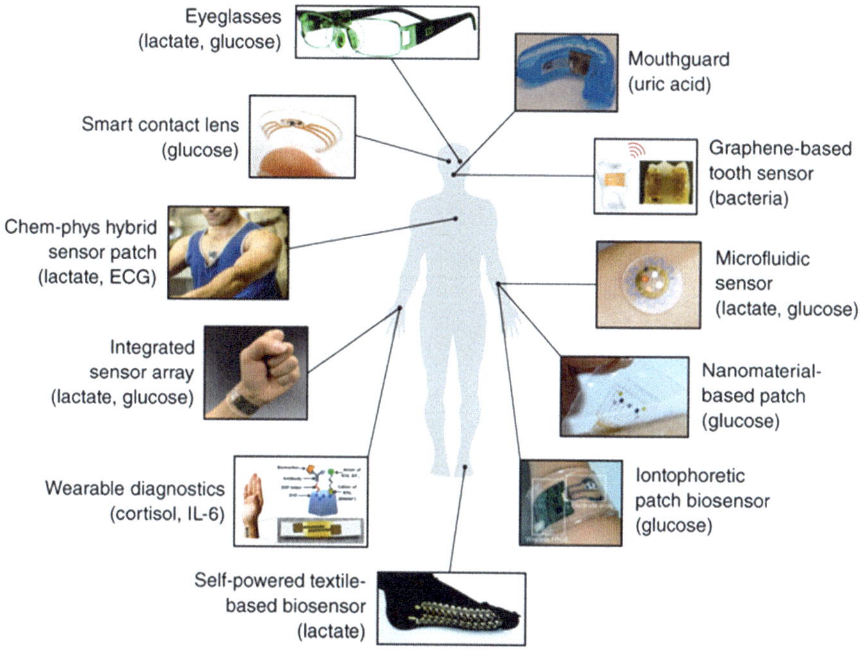

Fig. 9.10 Examples of wearable biosensors used in health care monitoring (Source: Kim et al. (2019))

for nearly 85% of the biosensors market (Rea et al. 2009) (Fig. 9.10). Biosensor-based techniques were implemented for detecting cytokines and antigen–antibody interactions. Hafnium oxide-based biosensor is a novel sensor which is used to detect human interleukin-10 (Lee et al. 2012) and human antigen by using electrochemical spectroscopy (Chen et al. 2010). Fluorometric immune affinity assay is used to detect pro- and anti-inflammatory cytokines in multiple organs failure syndromes (Caruso et al. 2010). The other technique used in the medical industry is early detection of cancer, and progression and response to treatment via biomarkers-based biosensors. Biosensors can also be used for the identification of pathogens. *Helicobacter pylori* are found in human adenoid tissue and tonsil; dental plaque; oral lesions, saliva, and stomach either as single cells (Neelapu et al. 2014) or biofilm (Challa and Neelapu 2018; Challa et al. 2018; Mohana Sheela et al. 2018; Neelapu et al. 2018; Surekha and Neelapu 2018). *H. pylori* were known for causing gastrointestinal disorders like gastritis, ulcers, and gastric cancer (Challa et al. 2019). Sometimes *H. pylori* may trigger some diseases like laryngitis and glossitis, pharyngitis, otitis, and sinusitis (Caruso et al. 2010). Many groups were successful in identifying new or alternative drug targets for the eradication of *H. pylori* infections (Neelapu 2018; Neelapu et al. 2013, 2015, 2016; Neelapu and Pavani 2013; Nammi et al. 2016; Nammi et al. 2017; Pasupuleti et al. 2017). Further, electrochemical biosensors were employed for early detection of *H. pylori* infection which may lead to gastrointestinal cancer (Chen et al. 2018) (Fig. 9.11). This will improve treatment

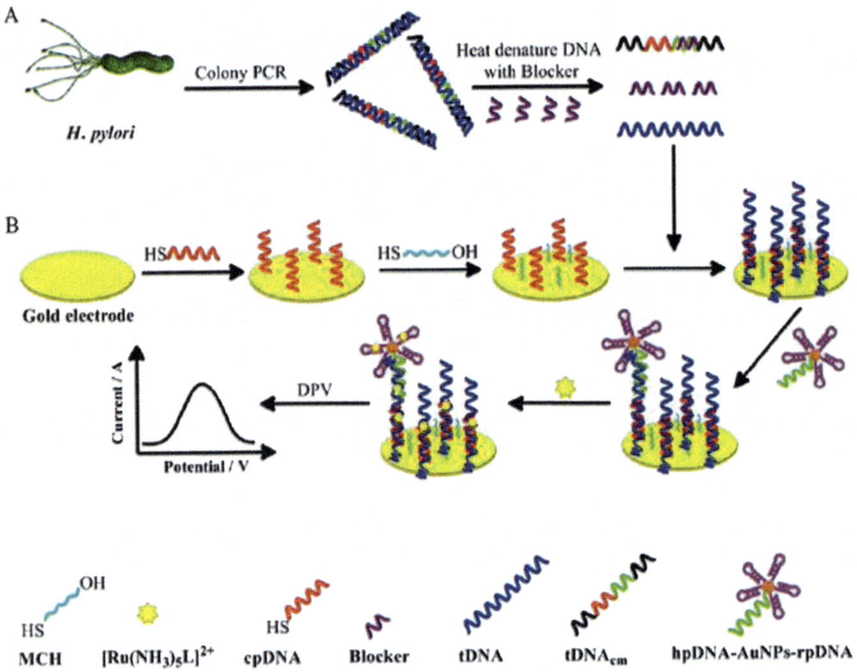

Fig. 9.11 Illustrations of (**a**) the amplification and acquirement of target dsDNA from *H. pylori* and releasing of the single strand tDNA from the dsDNA, and (**b**) the construction of the electrochemical sandwich DNA biosensor and the tDNA detection with the biosensor (Source: Li Chen et al. (2018))

for *H. pylori* infections leading to cancer. Thus, the above examples demonstrate the role of biosensors in the medical industry.

9.5.2 Biosensors Role in Drug Discovery

Identification of drugs and their targets for diseases like cancer, neurodegenerative diseases, cardiovascular diseases, infectious diseases is possible when biosensors are employed. Biosensors can detect protein localization, transcription, signal transduction, cell cycle, apoptosis, and gene expression (Morris 2013). Biosensors like fluorescence biosensors and FRET biosensors have an important role in the pharma industry. They also have a role in the discovery of targets and drugs for cancer and infectious diseases. Fluorescence biosensors identify targets, ions, metabolites, and biomarkers with high sensitivity and specificity, which can be detected and measured using fluorescent signals (Morris 2010). FRET biosensors play an important role as an imaging biomarker in diagnosing disease. Thus, the above examples demonstrate the role of biosensors in drug discovery.

9.5.3 Biosensors Role in Plant Studies

Plant science along with its traditional research methods uses new technologies like biosensors, sequencing, and molecular imaging to measure metabolites level, ions, etc. In plant models, it is very difficult to identify the locations of the event, transporters involved, and the participating receptors in the signaling event. Biosensors are used to get this key information successfully and also measure and visualize the actual process and signaling events. Professor Tsuen's research lab developed protein prototype biosensors to measure the calcium levels in live cells which are based on FRET (Tian et al. 2012). Other biosensors like chameleon biosensors help detect the missing components, regulation, transport, and metabolism of the analyte. Sugar sensors are based on fluorimeter-based biosensor which helps in recognizing sugar transporters when starved cells are exposed to glucose and also identifies genes that affect vacuolar or cytosol in yeast (Bermejo et al. 2010). Thus, biosensors play a very important role in plant research studies.

9.5.4 Biosensors Role in Fermentation Industries

Biosensors' role in fermentation industries to monitor the presence of products, by-products, and measure the conditions of a process is indispensable. This increases the efficacy, reproducibility with low and simple instrumentation at a low price, and easy methods. Some commercial biosensors are highly used in the fermentation industry to detect glucose, lactate, lysine, etc. Glucose biosensors are used to detect the glucose produced by enzymatic methods during the process of saccharification. Glutamate biosensors are used to detect the ion exchange of a glutamate supernatant during the process of glutamate production. Thus, the fermentation industry attracted biosensors with important roles (Yan et al. 2014).

9.5.5 Biosensors Role in Defense and Fighting Bioterrorism

Biological warfare agents like bacteria (*Bacillus anthracis*, *Brucella* sp., *Francisella tularensis*, *Yersinia pestis*) and bacterial toxins (*Staphylococcal* enterotoxin B, *Botulinum* toxin), and viruses (orthopoxviruses) are the typical biological weapons (Pohanka 2019). These biological weapons are either used by military or misused by a terrorist group to harm or kill humans. Biosensors help in detecting or identifying the biological weapons (organisms) like bacteria, virus, toxins, etc. that poses a threat as bioterrorism agents. For example, the prototypes, of optical biosensors, electrochemical biosensors, and piezoelectric biosensors, and commercially biosensors are available for assaying biological warfare agents (Pohanka 2019).

Thus, the role of biosensors attracted defense to use biosensors for fighting bioterrorism.

9.5.6 *Biosensors Role in Cell and its Metabolism*

Genetically, encoded biosensors monitor in vivo cellular metabolism by screening metabolites, regulation of gene expression, and mRNA regulatory domains. The three classes of biosensors like FRET biosensors, biosensors based on transcription factors, and the third class of biosensors like DNA biosensors are used to study and monitor in vivo cellular metabolism. FRET biosensors identify and screen metabolites that are responsible for cell metabolism. Biosensors based on transcription factors help in detecting the regulation of gene expression in response to changes in the environment and host. The third class of biosensors helps in detecting the mRNA regulatory domain and ribosomes in the bacterial systems (Berens and Suess 2015). Thus, biosensors can be employed in monitoring and understanding in vivo cellular metabolism.

9.5.7 *Biosensors Role in Ensuring Quality of Food*

Biosensors are capable of detecting the pathogens, pesticides, and artificial sweeteners in the food. Enzymatic biosensors and potentiometric biosensors are used in detecting bacterial cells in food, vegetables, and fruits (Torun et al. 2012). Screen-printed carbon electrode-based biosensor and flow-based biosensor detects pesticides in dairy products. Artificial sweeteners are extensively used in the food and cause severe problems like obesity, cardiovascular risks, and dental problems. Taste epithelium biosensors are used to detect the difference of taste in natural and artificial sweeteners (Zhang et al. 2014b). Thus, biosensors can be employed in identifying pathogens in contaminated food and adulterants in adulterated food (Fig. 9.12).

9.6 Role and Mechanism of Biosensor in Early Detection of Cancer

Cancers like breast, stomach, lung, colorectal, thyroid, renal, endometrial, pancreatic, and liver leukemia are life-threatening. Early diagnosis of cancer confirms the cancer stage and also prevents death due to cancer with appropriate treatment at the right times (National Institute Cancer 2015; Islam and Uddin 2017). Biomarkers and biosensors can help in early diagnosis or detection of cancers. The biomarkers identified using biosensors are listed in Table 9.3. Aptamers are used for detecting

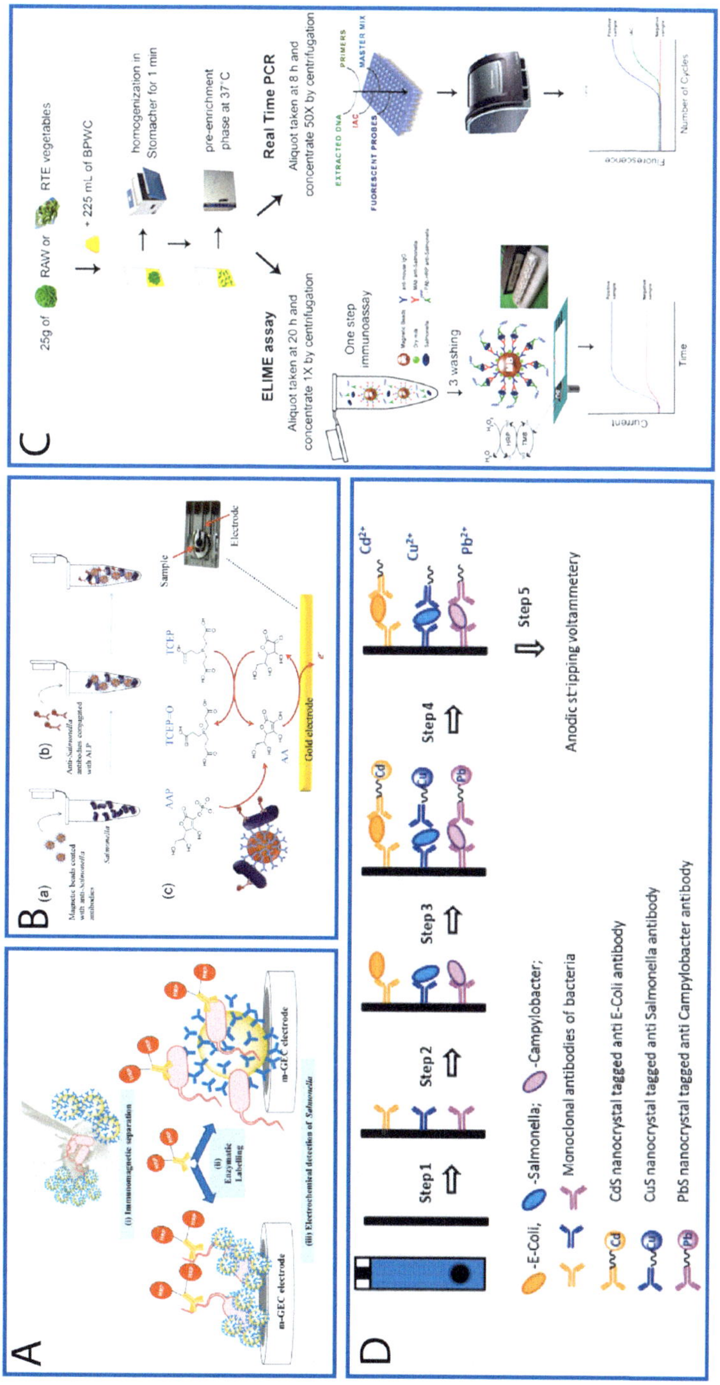

Fig. 9.12 (**a**) *Salmonella* detection based on electrochemical magneto-immunosensing, (**b**) Immunomagnetic pre-concentration and electrochemical detection based on redox cycling, (**c**) Salmonella detection with an ELIME (Enzyme-Linked-Immuno-Magnetic-Electrochemical)-based sandwich assay, (**d**) Multi-detection of pathogens using NC antibody conjugates and MWCNT-PAH/SPE (Source: Cinti et al. (2017))

Table 9.3 List of biomarkers identified using biosensor for cancer

S. no.	Biosensor types	Biomarker names	Cancer types	References
1.	Immunosensor	Indium tin oxide	Cancer	Canbaz and Sezgintürk (2014)
2.	Electrochemical	MGC-803	Gastric cancer	Zhang et al. (2014c)
3.	Electrochemical	Mir-106a	Gastric cancer	Daneshpour et al. (2016)
4.	Electrochemical	Pepsinogen I	Gastric cancer	Xie et al. (2015)
5.	Electrochemical	Cag A	Gastric cancer	Xie et al. (2015)
6.	Electrochemical	P53 oncoprotein	Gastric cancer	Xie et al. (2015)
7.	Electrochemical	Carbohydrate antigen 19-9	Gastric cancer	Xie et al. (2015)
8.	Electrochemical	Carcinoembryonic antigen	Gastric cancer	Xie et al. (2015)
9.	Optical	Apolipoprotein-A	Gastric cancer	Sciacca et al. (2013)
10.	Optical	Clusterin	Gastric cancer	Sciacca et al. (2013)
11.	Electrochemical	MiRNA-21	Gastric cancer	Li et al. (2016)
12.	Micro	Glucose	Gastric cancer	Wang et al. (2016)
13.	Amperometric	Hydrogen perioxide	Gastric cancer	Tabrizi et al. (2017)
14.	Electrochemical	Exosomes	Gastric cancer	Tabrizi et al. (2017)
15.	Graphene	Alpha fetoprotein	Liver cancer	Gu et al. (2019)
16.	Graphene	BRCA1, BRCA2	Breast cancer	Gu et al. (2019)
17.	Graphene	Somatostatin receptor subtype	Tumor	Feng et al. (2013)
18.	Electrochemical	Okadaic acid	Cancer	Eshghi et al. (2019)
19.	Aptamer	Prostate cancer antigen	Prostate cancer	Jolly et al. (2015)

specific targets that bind to it and cause conformational changes. The transducer converts these changes (signal) into response (Zhang et al. 2013). Based on the electrical response the electrochemical biosensor recognizes a hpRL-3 element specific to breast cancer (Asphahani and Zhang 2007). The point mutation in the p53 gene is detected using mass change biosensor. The mass change biosensor measures resonance frequency changes using biosensors with polymerase chain reactions. Calorimetric biosensors and graphene biosensors measure exothermic reactions and detect different types of cells in acute leukemia (Wang et al. 2017). The optical biosensors are helpful in monitoring changes in apoptosis of breast cancer cells. Optical biosensor changes based on fluorescence is used for effective diagnosis of throat cancer. Very specifically the change in wavelengths are measured by optical biosensors (Bohunicky and Mousa 2011). Immunosensors, enzyme-based biosensors help in identifying biomarkers based on antigen–antibody interactions and toxins that interact with the immune system (Wang 1998). Thus, biomarkers and biosensors can be employed for the detection of various cancers.

9.7 Role and Mechanism of Biosensor in Early Detection of Gastrointestinal Cancer

Gastrointestinal cancer is one of the most common cancers in the world and ranked fourth in the world Statistics show that gastrointestinal cancer is the second most related death in the world of cancer (Jemal et al. 2008). Gastrointestinal cancer needs an effective treatment and method for early diagnosis or detection (Bondy 2009). Early diagnosis or detection of gastrointestinal cancer is challenging, and there is a requirement to discover specific biomarkers. Major efforts have been made to develop techniques for biomarker detection and biosensors. Biosensors use DNA, antibody, antigen, enzyme, whole-cell, and cell organelle as a biological recognition element. The major focus is an electrochemical analysis of protein and metabolite biomarkers for potential prognosis and early detection of gastrointestinal cancer.

Volatile metabolites released by cancer cells are considered as essential indicators or biomarkers for cancer cell metabolism and biochemical process (Miekisch et al. 2004). These volatile compounds are very helpful in the early detection and diagnosis of gastrointestinal cancer (Phillips et al. 2010). Volatile biomarkers are coupled with electrochemical biosensors to detect gastrointestinal cancer. The identification of volatile biomarkers released from MGC-803 cancer cell lines is based on chromatogram patterns generated. Eight volatile metabolites (compounds) like 3-octanone; 4-isopropoxybutanol; 1,4-butanediol; nonanol; formic acid propyl ester; butanone; 1-butanol; 4-butoxy and dodecane, 2611-trimethy are identified in MGC-803 gastrointestinal cancer cell lines. The behaviors of electrochemical electrodes are observed by the well-defined peaks of MWNTs and Au-Ag nanocomposites. The electrochemical sensors detect the concentration of butanone and 3-octanone in the matrix of the anodic peak and negative shift potential. The detection of 3-octanone and butanone is based on the regression equation and detection of limitation. The electrochemical sensors were able to distinguish between the gastric mucous cells and the gastric cells. Thus, volatile biomarkers coupled with electrochemical biosensor are helpful in the early detection of gastrointestinal cancer.

Electrochemical biosensors coupled with protein biomarkers are used for early detection of gastrointestinal cancer. Protein biomarkers employed in the development of immunosensor are CA 72-4, interleukin-6, and CA-19-9 (Freitas et al. 2018). The methods that are employed for the detection of biomarkers are immunohistochemistry (IHC) and fluorescent in situ hybridization (FISH). These assays are sensitive and detect biomarkers in low concentrations in the sample. The conversion of recognition elements into a signal and transduction of this signal is based on electrochemical, piezoelectric, calorimetric, and optical sensors (Sassolas et al. 2012). Thus, volatile biomarkers and protein biomarkers coupled with electrochemical biosensor are helpful in the early detection of gastrointestinal cancer.

9.8 Conclusions and Future Perspectives

Cancer is a life-threatening disease and side effects are another problem when patients are subjected to treatment. To overcome this problem, biosensors with specific biomarkers are to be employed. Biosensors can provide key information on cancer for effective and safe treatment. Cost-effectiveness, reliability, accuracy, and less time consuming are important aspects of biosensors. Biosensors were already used in fields like drug discovery, fermentation industry, defense, food quality, environmental monitoring, metabolic studies, plant studies, etc. Now, biosensors emerged as the most powerful technologies for early detection and diagnosis of cancer. Biosensors improve the diagnostic capability by its sensitivity, specificity, reproducibility, linearity, and high-throughput screening. DNA, antibody, antigen, enzyme, whole-cell, and cell organelle are used as a biological recognition element for biosensors. The biological sample interacts with the element of the biosensor and forms a product. The product reaches the transducer, amplifies, records, and displays on the devices. The different types of biosensors are affinity biosensor, catalytic biosensor, metabolism biosensor, DNA biosensor, electrochemical biosensor, optical biosensor, mass change biosensor, graphene-based biosensor, amperometric biosensor, microbial biosensor, miRNA biosensor, and many more. Among them, the electrochemical biosensor was used for early detection of gastric cancer. Thus, it can be concluded that biosensors have an important role in early detection of gastrointestinal cancer.

Acknowledgments ND and NNRR are grateful to GITAM (Deemed-to-be-University) for providing necessary facilities to carry out the research work and for extending constant support.

Conflict of Interest The authors declare that there is no potential conflict of interest.

References

Afsahi S, Lerner MB, Goldstein JM, Lee J, Tang X, Bagarozzi DA Jr et al (2018) Novel graphene-based biosensor for early detection of Zika virus infection. Biosens Bioelectron 100:85–88

Altieri F, Di Stadio CS, Federico A, Miselli G, De Palma M, Rippa E, Arcari P (2017) Epigenetic alterations of gastrokine 1 gene expression in gastric cancer. Oncotarget 8(10):16899
Arora P, Sindhu A, Dilbaghi N, Chaudhury A (2011) Biosensors as innovative tools for the detection of food borne pathogens. Biosens Bioelectron 28(1):1–12
Asao T, Nagamachi Y, Fukuda T, Yazawa S (1991) Carcinoembryonic antigen levels in peritoneal washings can predict peritoneal recurrence after curative resection of gastric cancer. Cancer 68 (1):44–47
Asphahani F, Zhang M (2007) Cellular impedance biosensors for drug screening and toxin detection. Analyst 132(9):835–841
Aviel-Ronen S, Blackhall FH, Shepherd FA, Tsao MS (2006) K-ras mutations in non-small-cell lung carcinoma: a review. Clin Lung Cancer 8(1):30–38
Bahrami A, Khazaei A, Avan A (2018) Long non-coding RNA and microRNAs as novel potential biomarker and therapeutic target in the treatment of gastrointestinal cancers. Curr Pharm Des 24 (39):4599–4600
Berens C, Suess B (2015) Riboswitch engineering—making the all-important second and third steps. Curr Opin Biotechnol 31:10–15
Berge W, Mundt K, Luu H, Boffetta P (2018) Genital use of talc and risk of ovarian cancer: a meta-analysis. Eur J Cancer Prev 27(3):248–257
Bermejo C, Haerizadeh F, Takanaga H, Chermak D, Frommer WB (2010) Dynamic analysis of cytosolic glucose and ATP levels in yeast using optical sensors. Biochem J 432(2):399–406
Bermejo C, Haerizadeh F, Takanaga H, Chermak D, Frommer WB (2011) Optical sensors for measuring dynamic changes of cytosolic metabolite levels in yeast. Nat Protoc 6(11):1806
Bhalla N, Jolly P, Formisano N, Estrela P (2016) Introduction to biosensors. Essays Biochem 60 (1):1–8
Birgisson H, Nielsen HJ, Christensen IJ, Glimelius B, Brünner N (2010) Preoperative plasma TIMP-1 is an independent prognostic indicator in patients with primary colorectal cancer: a prospective validation study. Eur J Cancer 46(18):3323–3331
Bohunicky B, Mousa SA (2011) Biosensors: the new wave in cancer diagnosis. Nanotechnol Sci Appl 4:1–10
Bondy M (2009) Cancer epidemiology and prevention. JAMA 301(10):1070–1076
Canbaz MÇ, Sezgintürk MK (2014) Fabrication of a highly sensitive disposable immunosensor based on indium tin oxide substrates for cancer biomarker detection. Anal Biochem 446:9–18
Cancer Facts & Figures (2019) Atlanta: American Cancer Society. http://www.cancer.org/content/dam/cancer-org/research/cancer-facts-and-statistics/annual-cancer-facts-and-figures/2019/cancer-facts-and-figures-2019.pdf. Accessed 12 Oct 2019
Cancer in Ireland 1994–2016 with estimates for 2016–2018. In: Annual report of the national cancer registry. (n.d.). https://www.ncri.ie/publications/statistical-reports/cancer-ireland-1994-2016-estimates-2016-2018-annual-report-national. Accessed 12 Oct 2019
Caruso R, Trunfio S, Milazzo F, Campolo J, De Maria R, Colombo T et al (2010) Early expression of pro-and anti-inflammatory cytokines in left ventricular assist device recipients with multiple organ failure syndrome. ASAIO J 56(4):313–318
Challa S, Neelapu NRR (2018) Association between horizontal gene transfer and adaptation of gastric human pathogen *Helicobacter pylori* to the host. In: Villa TG, Viñas M (eds) Horizontal gene transfer. Springer Nature Switzerland AG, Basel, pp 257–267
Challa S, Mohana Sheela G, Neelapu NRR (2018) Understanding the bacterial biofilm resistance to antibiotics and immune evasion. In: Chari PVB (ed) Implication of quorum sensing system in biofilm formation and virulence. Springer Nature, Basel, pp 369–381
Challa S, Dutta T, Pallaval VBC, Neelapu NRR (2019) Quorum sensing and multidrug resistance mechanism in *Helicobacter pylori*. In: Chari PVB (ed) Implication of quorum sensing and biofilm formation in medicine, agriculture and food industry. Springer Nature, Basel, pp 1–21
Chan SH, Wu CW, Li AFY, Chi CW, Lin WC (2008) miR-21 microRNA expression in human gastric carcinomas and its clinical association. Anticancer Res 28(2A):907–911

Chatterjee SK, Zetter BR (2005) Cancer biomarkers: knowing the present and predicting the future. Future Oncol 1(1):37–50

Chen YW, Liu M, Kaneko T, McIntyre PC (2010) Atomic layer deposited hafnium oxide gate dielectrics for charge-based biosensors. Electrochem Solid-State Lett 13(3):G29–G32

Chen LL, Cui HF, Fan SF, Li ZY, Han SY, Ma X et al (2018) Detection of *Helicobacter pylori* in dental plaque using a DNA biosensor for noninvasive diagnosis. RSC Adv 8(38):21075–21083

Cheng F, Kardashliev T, Pitzler C, Shehzad A, Lue H, Bernhagen J et al (2015) A competitive flow cytometry screening system for directed evolution of therapeutic enzyme. ACS Synth Biol 4 (7):768–775

Cho WC, Yip TT, Yip C, Yip V, Thulasiraman V, Ngan RK et al (2004) Identification of serum amyloid a protein as a potentially useful biomarker to monitor relapse of nasopharyngeal cancer by serum proteomic profiling. Clin Cancer Res 10(1):43–52

Choi S, Chae J (2012) An array of microliter-sized microbial fuel cells generating 100 μW of power. Sensors Actuators A Phys 177:10–15

Chong H, Ching CB (2016) Development of colorimetric-based whole-cell biosensor for organophosphorus compounds by engineering transcription regulator DmpR. ACS Synth Biol 5 (11):1290–1298

Cinti S, Volpe G, Piermarini S, Delibato E, Palleschi G (2017) Electrochemical biosensors for rapid detection of foodborne *Salmonella*: a critical overview. Sensors 17(8):1910

Correia CN, Nalpas NC, McLoughlin KE, Browne JA, Gordon SV, MacHugh DE, Shaughnessy RG (2017) Circulating microRNAs as potential biomarkers of infectious disease. Front Immunol 8:118

Curea FG, Hebbar M, Ilie SM, Bacinschi XE, Trifanescu OG, Botnariuc I, Anghel RM (2017) Current targeted therapies in HER2-positive gastric adenocarcinoma. Cancer Biother Radiopharm 32(10):351–363

D'Souza SF (2001) Microbial biosensors. Biosens Bioelectron 16(6):337–353

Damborský P, Švitel J, Katrlík J (2016) Optical biosensors. Essays Biochem 60(1):91–100

Daneshpour M, Omidfar K, Ghanbarian H (2016) A novel electrochemical nanobiosensor for the ultrasensitive and specific detection of femtomolar-level gastric cancer biomarker miRNA-106a. Beilstein J Nanotechnol 7(1):2023–2036

Dhull V, Gahlaut A, Dilbaghi N, Hooda V (2013) Acetylcholinesterase biosensors for electrochemical detection of organophosphorus compounds: a review. Biochem Res Int

Duffy MJ (1998) CA 19-9 as a marker for gastrointestinal cancers: a review. Ann Clin Biochem 35 (3):364–370

Duffy MJ, O'Donovan N, Brennan DJ, Gallagher WM, Ryan BM (2007) Survivin: a promising tumor biomarker. Cancer Lett 249(1):49–60

Duffy MJ, Lamerz R, Haglund C, Nicolini A, Kalousova M, Holubec L, Sturgeon C (2014) Tumor markers in colorectal cancer, gastric cancer and gastrointestinal stromal cancers: European group on tumor markers 2014 guidelines update. Int J Cancer 134(11):2513–2522

Elimova E, Wadhwa R, Shiozaki H, Sudo K, Estrella JS, Badgwell BD et al (2015) Molecular biomarkers in gastric cancer. J Natl Compr Cancer Netw 13(4):e19–e29

Eshghi MG, Fruhideh Z, Alviri VM, Asem MM (2019) Electrochemical biosensors for cancer detection using different biomarkers. In: 2019 IEEE 9th Annual Computing and Communication Workshop and Conference (CCWC), Las Vegas, pp 0989–0996

Espinosa JR, Galván M, Quiñones AS, Ayala JL, Durón SM (2019) DNA biosensor based on double-layer discharge for the detection of HPV type 16. Sensors 19(18):3956

Feng B, Guo L, Wang L, Li F, Lu J, Gao J et al (2013) A graphene oxide-based fluorescent biosensor for the analysis of peptide–receptor interactions and imaging in somatostatin receptor subtype 2 overexpressed tumor cells. Anal Chem 85(16):7732–7737

Foa PAOLO, Fornier M, Miceli ROSALBA, Seregni E, Santambrogio L, Nosotti M et al (1999) Tumour markers CEA, NSE, SCC, TPA and CYFRA 21.1 in resectable non-small cell lung cancer. Anticancer Res 19(4C):3613–3618

Frazao CR, Maton V, François JM, Walther T (2018) Development of a metabolite sensor for high-throughput detection of aldehydes in Escherichia coli. Front Bioeng Biotechnol 6:118

Freitas M, Nouws HP, Delerue-Matos C (2018) Electrochemical biosensing in cancer diagnostics and follow-up. Electroanalysis 30(8):1584–1603

Fukushige S, Horii A (2013) DNA methylation in cancer: a gene silencing mechanism and the clinical potential of its biomarkers. Tohoku J Exp Med 229(3):173–185

García-Alonso J, Greenway GM, Hardege JD, Haswell SJ (2009) A prototype microfluidic chip using fluorescent yeast for detection of toxic compounds. Biosens Bioelectron 24(5):1508–1511

Ghotaslou R, Leylabadlo HE, Nasiri MJ, Dabiri H, Hashemi A (2018) Risk of gastric cancer in association with *Helicobacter pylori* different virulence factors: a systematic review and meta-analysis. Microb Pathog 118:214–219

Goetsch CM (2011) Genetic tumor profiling and genetically targeted cancer therapy. In: Seminars in oncology nursing, vol 27. WB Saunders, Philadelphia, pp 34–44

Gold P, Freedman SO (1965a) Demonstration of tumor-specific antigens in human colonic carcinomata by immunological tolerance and absorption techniques. J Exp Med 121(3):439–462

Gold P, Freedman SO (1965b) Specific carcinoembyronic antigens of the human digestive system. J Exp Med 122:467

Gómez-Cebrián N, Rojas-Benedicto A, Albors-Vaquer A, López-Guerrero JA, Pineda-Lucena A, Puchades-Carrasco L (2019) Metabolomics contributions to the discovery of prostate cancer biomarkers. Metabolites 9(3):48

Gu H, Tang H, Xiong P, Zhou Z (2019) Biomarkers-based biosensing and bioimaging with graphene for cancer diagnosis. Nanomaterials 9(1):130

Hájková A, Barek J, Vyskočil V (2017) Electrochemical DNA biosensor for detection of DNA damage induced by hydroxyl radicals. Bioelectrochemistry 116:1–9

Heo SH, Lee SJ, Ryoo HM, Park JY, Cho JY (2007) Identification of putative serum glycoprotein biomarkers for human lung adenocarcinoma by multilectin affinity chromatography and LC-MS/MS. Proteomics 7(23):4292–4302

Islam MT, Uddin MA (2017) Biosensors, the emerging tools in the identification and detection of cancer markers. J Gynecol Womens Health 5(4):555667

Jainish P, Prittesh P (2017) Biosensors and biomarkers: promising tools for cancer diagnosis. Int J Biosens Bioelectron 3(4):00072

Jemal A, Siegel R, Ward E, Hao Y, Xu J, Murray T, Thun MJ (2008) Cancer statistics, 2008. CA Cancer J Clin 58(2):71–96

Jolly P, Formisano N, Estrela P (2015) DNA aptamer-based detection of prostate cancer. Chem Pap 69(1):77–89

Kavita V (2017) DNA biosensors—a review. J Bioeng Biomed Sci 7:2

Kdadra M, Höckner S, Leung H, Kremer W, Schiffer E (2019) Metabolomics biomarkers of prostate cancer: a systematic review. Diagnostics 9(1):21

Kim JH, Skates SJ, Uede T, Wong KK, Schorge JO, Feltmate CM et al (2002) Osteopontin as a potential diagnostic biomarker for ovarian cancer. JAMA 287(13):1671–1679

Kim J, Campbell AS, de Ávila BEF, Wang J (2019) Wearable biosensors for healthcare monitoring. Nat Biotechnol 37(4):389–406

Kumar J, Jha SK, D'Souza SF (2006) Optical microbial biosensor for detection of methyl parathion pesticide using *Flavobacterium* sp. whole cells adsorbed on glass fiber filters as disposable biocomponent. Biosens Bioelectron 21(11):2100–2105

Kwon KK, Lee DH, Kim SJ, Choi SL, Rha E, Yeom SJ et al (2018) Evolution of enzymes with new specificity by high-throughput screening using DmpR-based genetic circuits and multiple flow cytometry rounds. Sci Rep 8(1):2659

Lee M, Zine N, Baraket A, Zabala M, Campabadal F, Caruso R et al (2012) A novel biosensor based on hafnium oxide: application for early stage detection of human interleukin-10. Sensors Actuators B Chem 175:201–207

Leech D (1994) Affinity biosensors. Chem Soc Rev 23(3):205–213

Lei Y, Chen W, Mulchandani A (2006) Microbial biosensors. Anal Chim Acta 568(1-2):200–210

Li Q, Shao Y, Zhang X, Zheng T, Miao M, Qin L et al (2015a) Plasma long noncoding RNA protected by exosomes as a potential stable biomarker for gastric cancer. Tumor Biol 36 (3):2007–2012

Li P, Chen S, Chen H, Mo X, Li T, Shao Y et al (2015b) Using circular RNA as a novel type of biomarker in the screening of gastric cancer. Clin Chim Acta 444:132–136

Li B, Liu F, Peng Y, Zhou Y, Fan W, Yin H et al (2016) Two-stage cyclic enzymatic amplification method for ultrasensitive electrochemical assay of microRNA-21 in the blood serum of gastric cancer patients. Biosens Bioelectron 79:307–312

Liu Y, Liu Y, Wang M (2017) Design, optimization and application of small molecule biosensor in metabolic engineering. Front Microbiol 8:2012

Maciel CM, Junqueira M, Paschoal MEM, Kawamura MT, Duarte RLM, Carvalho MDGDC, Domont GB (2005) Differential proteomic serum pattern of low molecular weight proteins expressed by adenocarcinoma lung cancer patients. J Exp Ther Oncol 5(1):31–38

Maeda K, Chung YS, Takatsuka S, Ogawa Y, Sawada T, Yamashita Y et al (1995) Tumor angiogenesis as a predictor of recurrence in gastric carcinoma. J Clin Oncol 13(2):477–481

Magklara A, Scorilas A, Catalona WJ, Diamandis EP (1999) The combination of human glandular kallikrein and free prostate-specific antigen (PSA) enhances discrimination between prostate cancer and benign prostatic hyperplasia in patients with moderately increased total PSA. Clin Chem 45(11):1960–1966

Malhotra S, Verma A, Tyagi N, Kumar V (2017) Biosensors: principle, types and applications. Int J Adv Res Innov Ideas Educ 3(2):3639–3644

Marie Vedeld H, Andresen K, Andrassy Eilertsen I, Nesbakken A, Seruca R, Gladhaug IP et al (2015) The novel colorectal cancer biomarkers CDO1, ZSCAN18 and ZNF331 are frequently methylated across gastrointestinal cancers. Int J Cancer 136(4):844–853

Matsuoka T, Yashiro M (2018) Biomarkers of gastric cancer: current topics and future perspective. World J Gastroenterol 24(26):2818

Medley CD, Smith JE, Tang Z, Wu Y, Bamrungsap S, Tan W (2008) Gold nanoparticle-based colorimetric assay for the direct detection of cancerous cells. Anal Chem 80(4):1067–1072

Miekisch W, Schubert JK, Noeldge-Schomburg GF (2004) Diagnostic potential of breath analysis—focus on volatile organic compounds. Clin Chim Acta 347(1–2):25–39

Mohana Sheela G, Prathyusha AMVN, Neelapu NRR, Bramhachari PV (2018) Intra and inter-species communication in microbes: living with complex and sociable neighbors. In: Chari PVB (ed) Implication of quorum sensing system in biofilm formation and virulence. Springer Nature, Basel, pp 7–16

Mohsin M, Abdin MZ, Nischal L, Kardam H, Ahmad A (2013) Genetically encoded FRET-based nanosensor for in vivo measurement of leucine. Biosens Bioelectron 50:72–77

Morgan SA, Nadler DC, Yokoo R, Savage DF (2016) Biofuel metabolic engineering with biosensors. Curr Opin Chem Biol 35:150–158

Morris MC (2010) Fluorescent biosensors of intracellular targets from genetically encoded reporters to modular polypeptide probes. Cell Biochem Biophys 56(1):19

Morris MC (2013) Fluorescent biosensors—probing protein kinase function in cancer and drug discovery. Biochim Biophys Acta 1834(7):1387–1395

Nammi D, Srimath-Tirumala-Peddinti RCPK, Neelapu NRR (2016) Identification of drug targets in *Helicobacter pylori* by in silico analysis: possible therapeutic implications for gastric cancer. Curr Cancer Drug Targets 16:79–98

Nammi D, Yarla NS, Chubarev VN, Tarasov VV, Barreto GE, Pasupulati CAM, Aliev G, Neelapu NRR (2017) A systematic in-silico analysis of *Helicobacter pylori* pathogenic islands for identification of novel drug target candidates. Curr Genomics 18:450–465

National Institute Cancer (2015) What is cancer? https://www.cancer.gov/about-cancer/understanding/what-is-cancer. Accessed 12 Oct 2019

Neelapu RR (2018) Role and regulation of transcriptional factors in gastric cancer. In: Role of transcription factors in gastrointestinal malignancies. Springer Nature, Hiedelberg, pp 107–130. isbn:978-981-10-6727-3

Neelapu NRR, Pavani T (2013) Identification of novel drug targets in HpB38, HpP12, HpG27, Hpshi470, HpSJM180 strains of *Helicobacter pylori*: an *in silico* approach for therapeutic intervention. Curr Drug Targets 14:601–611

Neelapu NRR, Srimath-Tirumala-Peddinti RCPK, Nammi D, Pasupuleti ACM (2013) New strategies and paradigm for drug target discovery: a special focus on infectious diseases tuberculosis, malaria, leishmaniasis, trypanosomiasis and gastritis. Infect Disorder Drug Targets 13 (5):352–364

Neelapu NRR, Nammi D, Pasupuleti ACM, Surekha C (2014) *Helicobacter pylori* induced gastric inflammation, ulcer, and cancer: a pathogenesis perspective. Interdiscip J Microinflammation 1:113

Neelapu NRR, Mutha NVR, Akula S (2015) Identification of potential drug targets in *Helicobacter pylori* strain HPAG1 by *in silico* genome analysis. Infect Disord Drug Targets 15:106–117

Neelapu NRR, Nammi D, Pasupuleti AMC, Challa S (2016) Targets against *Helicobacter pylori* and other tumor-producing bacteria. In: New weapons to control bacterial growth. Springer, Cham, pp 239–279

Neelapu NRR, Titash D, Surekha C (2018) Quorum sensing and its role in Agrobacterium mediated gene transfer. In: Chari PVB (ed) Implication of quorum sensing system in biofilm formation and virulence. Springer Nature, Basel, pp 259–275

Nesterova M, Johnson N, Cheadle C, Cho-Chung YS (2006) Autoantibody biomarker opens a new gateway for cancer diagnosis. Biochim Biophys Acta 1762(4):398–403

Niloff JM, Knapp RC, Schaetzl E, Reynolds CAROL, Bast JR (1984) CA125 antigen levels in obstetric and gynecologic patients. Obstet Gynecol 64(5):703–707

Okayama H, Kumamoto K, Saitou K, Hayase S, Kofunato Y, Sato Y et al (2009) CD44v6, MMP-7 and nuclear Cdx2 are significant biomarkers for prediction of lymph node metastasis in primary gastric cancer. Oncol Rep 22(4):745–755

Oue N, Sentani K, Sakamoto N, Yasui W (2015) Clinicopathologic and molecular characteristics of gastric cancer showing gastric and intestinal mucin phenotype. Cancer Sci 106(8):951–958

Paliouras M, Borgono C, Diamandis EP (2007) Human tissue kallikreins: the cancer biomarker family. Cancer Lett 249(1):61–79

Pasupuleti AMP, Nammi D, Neelapu NRR (2017) Screening and identification of drug targets and vaccine candidates for *Helicobacter pylori* strain Hp26695. Int J Rect Sci Res 8 (4):16384–16395

Peroza EA, Ewald JC, Parakkal G, Skotheim JM, Zamboni N (2015) A genetically encoded förster resonance energy transfer sensor for monitoring in vivo trehalose-6-phosphate dynamics. Anal Biochem 474:1–7

Phillips M, Cataneo RN, Saunders C, Hope P, Schmitt P, Wai J (2010) Volatile biomarkers in the breath of women with breast cancer. J Breath Res 4(2):026003

Planque C, Li L, Zheng Y, Soosaipillai A, Reckamp K, Chia D et al (2008) A multiparametric serum kallikrein panel for diagnosis of non–small cell lung carcinoma. Clin Cancer Res 14 (5):1355–1362

Pohanka M (2019) Current trends in the biosensors for biological warfare agents assay. Materials (Basel) 12(14):2303. https://doi.org/10.3390/ma12142303

Rai AJ, Zhang Z, Rosenzweig J, Shih IM, Pham T, Fung ET et al (2002) Proteomic approaches to tumor marker discovery: identification of biomarkers for ovarian cancer. Arch Pathol Lab Med 126(12):1518–1526

Ray S, Reddy PJ, Jain R, Gollapalli K, Moiyadi A, Srivastava S (2011) Proteomic technologies for the identification of disease biomarkers in serum: advances and challenges ahead. Proteomics 11 (11):2139–2161

Rea G, Polticelli F, Antonacci A, Scognamiglio V, Katiyar P, Kulkarni SA et al (2009) Structure-based design of novel Chlamydomonas reinhardtii D1-D2 photosynthetic proteins for herbicide monitoring. Protein Sci 18(10):2139–2151

Rice M, Stoyanova T (2018) Biomarkers for diagnosis and prognosis of prostate cancer. In: Prostatectomy. IntechOpen

Rogers JK, Taylor ND, Church GM (2016) Biosensor-based engineering of biosynthetic pathways. Curr Opin Biotechnol 42:84–91

Saju P, Murata-Kamiya N, Hayashi T, Senda Y, Nagase L, Noda S et al (2016) Host SHP1 phosphatase antagonizes *Helicobacter pylori* CagA and can be downregulated by Epstein–Barr virus. Nat Microbiol 1(4):16026

Salazar R, Roepman P, Capella G, Moreno V, Simon I, Dreezen C et al (2011) Gene expression signature to improve prognosis prediction of stage II and III colorectal cancer. J Clin Oncol 29 (1):17–24

Saleem M, Adhami VM, Zhong W, Longley BJ, Lin CY, Dickson RB et al (2006) A novel biomarker for staging human prostate adenocarcinoma: overexpression of matriptase with concomitant loss of its inhibitor, hepatocyte growth factor activator inhibitor-1. Cancer Epidemiol Prev Biomark 15(2):217–227

Sassolas A, Prieto-Simón B, Marty JL (2012) Biosensors for pesticide detection: new trends. Am J Anal Chem 3(03):210

Sciacca B, François A, Hoffmann P, Monro TM (2013) Multiplexing of radiative-surface plasmon resonance for the detection of gastric cancer biomarkers in a single optical fiber. Sensors Actuators B Chem 183:454–458

Scognamiglio V, Pezzotti G, Pezzotti I, Cano J, Buonasera K, Giannini D, Giardi MT (2010) Biosensors for effective environmental and agrifood protection and commercialization: from research to market. Microchim Acta 170(3-4):215–225

Shih IM, Salani R, Fiegl M, Wang TL, Soosaipillai A, Marth C et al (2007) Ovarian cancer specific kallikrein profile in effusions. Gynecol Oncol 105(2):501–507

Song MY, Pan KF, Su HJ, Zhang L, Ma JL, Li JY et al (2012) Identification of serum microRNAs as novel non-invasive biomarkers for early detection of gastric cancer. PLoS Onc 7(3):e33608

Steinberg SM, Poziomek EJ, Engelmann WH, Rogers KR (1995) A review of environmental applications of bioluminescence measurements. Chemosphere 30(11):2155–2197

Su L, Jia W, Hou C, Lei Y (2011) Microbial biosensors: a review. Biosens Bioelectron 26 (5):1788–1799

Sun H, Wu L, Wei W, Qu X (2013) Recent advances in graphene quantum dots for sensing. Mater Today 16(11):433–442

Sung HJ, Cho JY (2008) Biomarkers for the lung cancer diagnosis and their advances in proteomics. BMB Rep 41(9):615–625

Surekha C, Neelapu NRR (2018) Quorum sensing in *Helicobacter pylori*: role of biofilm and its implications for antibiotic resistance and immune evasion. In: Chari PVB (ed) Implication of quorum sensing system in biofilm formation and virulence. Springer Nature, Basel, pp 361–381

Szunerits S, Boukherroub R (2018) Graphene-based biosensors. Interface focus 8(3):20160132

Tabrizi MA, Shamsipur M, Saber R, Sarkar S, Sherkatkhameneh N (2017) Flow injection amperometric sandwich-type electrochemical aptasensor for the determination of adenocarcinoma gastric cancer cell using aptamer-Au@ Ag nanoparticles as labeled aptamer. Electrochim Acta 246:1147–1154

Tian L, Hires SA, Looger LL (2012) Imaging neuronal activity with genetically encoded calcium indicators. Cold Spring Harbor Protocols 2012(6), pdb-top069609

Tiquia-Arashiro SM (2014) Biotechnological applications of thermophilic carboxydotrophs. In: Thermophilic carboxydotrophs and their applications in biotechnology. Springer, Cham, pp 29–101

Torun Ö, Boyacı İH, Temür E, Tamer U (2012) Comparison of sensing strategies in SPR biosensor for rapid and sensitive enumeration of bacteria. Biosens Bioelectron 37(1):53–60

Tothill IE (2009) Biosensors for cancer markers diagnosis. Semin Cell Dev Biol 20(1):55–62

Van Dorst B, Mehta J, Bekaert K, Rouah-Martin E, De Coen W, Dubruel P et al (2010) Recent advances in recognition elements of food and environmental biosensors: a review. Biosens Bioelectron 26(4):1178–1194

Wang J (1998) DNA biosensors based on peptide nucleic acid (PNA) recognition layers. A review. Biosens Bioelectron 13(7-8):757–762

Wang DG, Chen G, Wen XY, Wang D, Cheng ZH, Sun SQ (2015) Identification of biomarkers for diagnosis of gastric cancer by bioinformatics. Asian Pac J Cancer Prev 16(4):1361–1365

Wang X, Ma Y, Zhao M, Zhou M, Xiao Y, Sun Z, Tong L (2016) Determination of glucose in human stomach cancer cell extracts and single cells by capillary electrophoresis with a micro-biosensor. J Chromatogr A 1469:128–134

Wang L, Xiong Q, Xiao F, Duan H (2017) 2D nanomaterials based electrochemical biosensors for cancer diagnosis. Biosens Bioelectron 89:136–151

Warren JD, Xiong W, Bunker AM, Vaughn CP, Furtado LV, Roberts WL et al (2011) Septin 9 methylated DNA is a sensitive and specific blood test for colorectal cancer. BMC Med 9 (1):133

Watanabe Y, Kim HS, Castoro RJ, Chung W, Estecio MR, Kondo K et al (2009) Sensitive and specific detection of early gastric cancer with DNA methylation analysis of gastric washes. Gastroenterology 136(7):2149–2158

Wu CC, Chen HC, Chen SJ, Liu HP, Hsieh YY, Yu CJ et al (2008) Identification of collapsin response mediator protein-2 as a potential marker of colorectal carcinoma by comparative analysis of cancer cell secretomes. Proteomics 8(2):316–332

Xie Y, Zhi X, Su H, Wang K, Yan Z, He N et al (2015) A novel electrochemical microfluidic chip combined with multiple biomarkers for early diagnosis of gastric cancer. Nanoscale Res Lett 10 (1):477

Yakabe T, Nakafusa Y, Sumi K, Miyoshi A, Kitajima Y, Sato S et al (2010) Clinical significance of CEA and CA19-9 in postoperative follow-up of colorectal cancer. Ann Surg Oncol 17 (9):2349–2356

Yamamoto H, Watanabe Y, Oikawa R, Morita R, Yoshida Y, Maehata T et al (2016) BARHL2 methylation using gastric wash DNA or gastric juice exosomal DNA is a useful marker for early detection of gastric cancer in an *H. pylori*-independent manner. Clin Transl Gastroenterol 7(7): e184

Yan C, Dong F, Chun-yuan B, Si-rong Z, Jian-guo S (2014) Recent progress of commercially available biosensors in China and their applications in fermentation processes. J Northeast Agric Univ 21(4):73–85

Yasui W, Oue N, Ito R, Kuraoka K, Nakayama H (2004) Search for new biomarkers of gastric cancer through serial analysis of gene expression and its clinical implications. Cancer Sci 95 (5):385–392

Yilmaz A, Ece F, Bayramgürler B, Akkaya E, Baran R (2001) The value of CA125 in the evaluation of tuberculosis activity. Respir Med 95(8):666–669

Yuan CL, Li H, Zhu L, Liu Z, Zhou J, Shu Y (2016) Aberrant expression of long noncoding RNA PVT1 and its diagnostic and prognostic significance in patients with gastric cancer. Neoplasma 63(3):442–449

Zagorodko O, Spadavecchia J, Serrano AY, Larroulet I, Pesquera A, Zurutuza A et al (2014) Highly sensitive detection of DNA hybridization on commercialized graphene-coated surface plasmon resonance interfaces. Anal Chem 86(22):11211–11216

Zamay TN, Zamay GS, Kolovskaya OS, Zukov RA, Petrova MM, Gargaun A et al (2017) Current and prospective protein biomarkers of lung cancer. Cancers 9(11):155

Zhang Z, Bast RC, Yu Y, Li J, Sokoll LJ, Rai AJ et al (2004) Three biomarkers identified from serum proteomic analysis for the detection of early stage ovarian cancer. Cancer Res 64 (16):5882–5890

Zhang Y, Yang D, Weng L, Wang L (2013) Early lung cancer diagnosis by biosensors. Int J Mol Sci 14(8):15479–15509

Zhang Y, Gao G, Liu H, Fu H, Fan J, Wang K, Chen Y, Li B, Zhang C, Zhi X, He L (2014a) Identification of volatile biomarkers of gastric cancer cells and ultrasensitive electrochemical detection based on sensing interface of Au-Ag alloy coated MWCNTs. Theranostics 4(2):154

Zhang F, Zhang Q, Zhang D, Lu Y, Liu Q, Wang P (2014b) Biosensor analysis of natural and artificial sweeteners in intact taste epithelium. Biosens Bioelectron 54:385–392

Zhang Y, Gao G, Liu H, Fu H, Fan J, Wang K et al (2014c) Identification of volatile biomarkers of gastric cancer cells and ultrasensitive electrochemical detection based on sensing interface of Au-Ag alloy coated MWCNTs. Theranostics 4(2):154

Zheng Y, Katsaros D, Shan SJ, De La Longrais IR, Porpiglia M, Scorilas A et al (2007) A multiparametric panel for ovarian cancer diagnosis, prognosis, and response to chemotherapy. Clin Cancer Res 13(23):6984–6992

Chapter 10
Application of Nanotechnology in Early Detection of Gastrointestinal Cancer

Nageswara Rao Reddy Neelapu and Deepthi Nammi

Abstract Cancer biomarkers play an essential role in the diagnosis and prediction of cancer to improve treatment efficacy in gastrointestinal cancer patients. Promising technologies can be used to meet the current challenges and nanotechnology is a powerful technique for early detection of gastrointestinal cancer. Biomarkers also improve cancer treatment with fewer side effects and increase the patient's lifetime. This chapter discusses the need for nanobiosensors, nanomaterials used for biosensors, nanobiosensors, and its applications. This provides information and insights on the role of nanotechnology in the early detection of gastrointestinal cancer.

Keywords Biomarkers · Biosensors · Cancer · Early detection of cancer · Gastrointestinal cancer · Nanobiosensors

10.1 Introduction

Biological and chemical sensory points which are used to convey essential information to the world are nanosensors. A sensor is an important device that responds to physical stimulus and converts this stimulus into a measurable quantity by data acquisition. Nanotechnology's role has increased in the detection of tumor-specific biomarkers and become a very promising technology to detect cancer earlier. This earlier detection of cancer improves the survival of the patients in the long term, as well it helps in improving the therapeutic outcome and life quality of the patient (Salvati et al. 2015). Various cancers like lung, prostate, colorectal, and breast are widespread cancers in the world (Peng et al. 2010). Biomarkers are detected using novel nanosensors, and biomarkers are identified based on selectivity and sensitivity

N. R. R. Neelapu · D. Nammi (✉)
Department of Biochemistry and Bioinformatics, GITAM Institute of Science, Gandhi Institute of Technology and Management (GITAM) (Deemed-to-be-University), Visakhapatnam, Andhra Pradesh, India

P. Veera Bramhachari, N. R. R. Neelapu (eds.), *Recent Advancements in Biomarkers and Early Detection of Gastrointestinal Cancers*, Diagnostics and Therapeutic Advances in GI Malignancies, https://doi.org/10.1007/978-981-15-4431-6_10

(Dasilva et al. 2012). Some of the popular nanomaterials are carbon nanotubes, quantum dots, and gold nanoparticles. The early detection of cancer and malignant tumors from biological fluids using nanowires, nano cantilever is very essential (Ferrari 2005). Nanovectors and imaging moieties are two subfields to oncology. The advantage of these is multifunctionality for cancer-specific therapeutics, and imaging agents used for therapeutic efficacy by thousands of nanovectors as biomarker detection agents (Ferrari 2005). The detection of cancer is complicated due to its similarity between healthy and diseased tissue (Hanahan and Weinberg 2000). Early detection of cancer is very essential to eradicate cancer-related deaths worldwide (Higgins and Ettinger 2009). Gastrointestinal cancer is one of the most common and leading cancer worldwide, and it is very difficult to detect early stage (Bondy 2009). Therefore, it is challenging and there is a need to establish a method for early detection of gastrointestinal cancer (Cui et al. 2005). The major drawback for early detection of cancer is the lack of specific biomarkers for early stages of cancer. Hence, there is a need for discovering new biomarkers for the early stages of cancer. Detection of biomarkers at an early stage of cancer and also when biomarkers are in trace amounts is possible with biosensors, especially with nanobiosensors (Miekisch et al. 2004). Nanobiosensors help in identifying early cancer cells, i.e., by analysis of biomarkers like volatile metabolites. This indicates the potential progress in the detection of cancer cells for diagnosis and early warning of gastrointestinal cancer (Phillips et al. 2010). This chapter in detail discusses nanobiosensors, nanomaterials used for biosensors, nanosensors, and its applications, and the role of nanobiosensors in early detection of gastrointestinal cancer.

10.1.1 Need for Use of Nanobiosensors

Nanobiosensors are one of the amazing new era devices for diagnosis of disease. Nanobiosensors measure biological events through a compact probe using various detection technologies. Current technology creates a new set of nanobiosensors for disease diagnosis. Protein or DNA sequences can be detected or quantified by a device known as a biosensor. Many biosensors use an immobilized probe that binds the molecules selectively and the molecule is sensed by detecting a localized surface change. These changes can be measured using various methods like surface plasmon fluorescence or resonance, magnetic particles, resonant cantilever, etc. The label-free sensor is a well-designed sensor that widely uses biosensors for the detection of specific molecules. Some biosensors detect biomarkers using current or voltage measurement which is widely used due to cost-effectiveness, less power, and low error rate. Many sensors are designed very effectively for a diagnostic point of view, and these kinds of sensors play a very prominent role as point-of-care disease diagnostics.

10.1.2 Need for Real-Time Measurements

Biomolecules are very helpful to determine the disease condition state, and these biomolecules are known as biomarkers. Effective analysis or diagnosis needs real-time measurements of biological analytes, playing an essential role in data generation, processing, decision-making, rapid manipulation, etc. (Prasad 2014). To handle multiple requirements, there is a need for multiscale biosensors to monitor-specific analytes at low concentrations from a different and wide range of environments. So, the integration of nanomaterials (microfluidic approach) with the semiconductor industry (integrated circuits) helps in fluid manipulation, separation, and detection techniques. Therefore, there is a need to focus on real-time measurements chip to detect and also construct best and effective nanobiosensors with multifunctions (Reddy 2007).

10.1.3 Biosensors and Nanobiosensors

A biosensor is a device with a biologically active element and physical transducer to measure signal proportional to the concentration of the sample, and the response changed to signal can be measured electronically, optically, electrochemically, mechanically, calorimetrically, etc. Biosensors are classified according to signal transduction and biorecognition elements. Biosensors play a very prominent role on a large scale and are very beneficial for many commercial applications (Arnold and Meyerhoff 1988). The major advances in biosensors are creating new nanobiosensors using nanotechnologies. These devices measure the biological events by using optical, electronic, and magnetic technology by the compact probe. These devices will change and conquer the new era in nanobiotechnology for early detection and diagnosis of diseases (Di Giusto et al. 2005).

10.2 Nanomaterials for Biosensors

Literature reports quantum dots; carbon nanotubes; nanopores and nanorods; nanowires and cantilevers; and nanoparticles as nanomaterials used for the development of nanobiosensors (Fig. 10.1).

10.2.1 Quantum Dots

Quantum dots are composed of semiconductors which are light-emitting nanocrystals (size 2-nm) with a wide brand adsorption display and emission bands

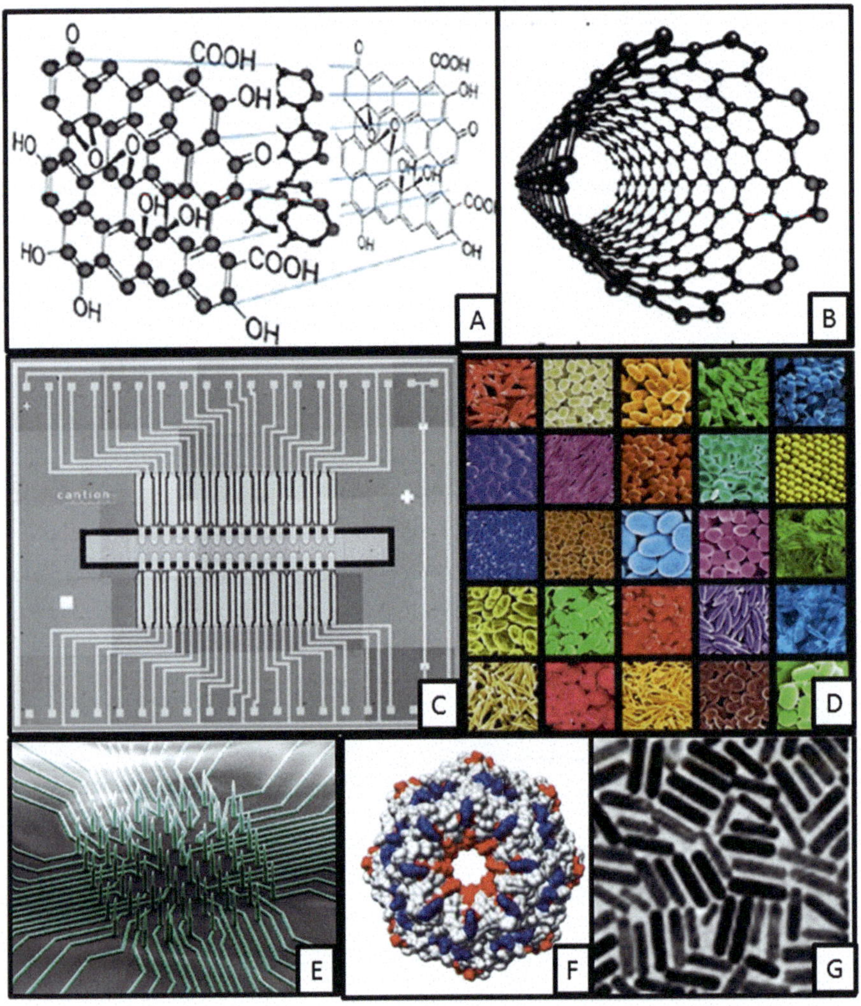

Fig. 10.1 Structure of nanomaterials (**a**) carbon dots (Source: Demchenko and Dekaliuk (2013)); (**b**) carbon nanotube (Source: Demchenko and Dekaliuk (2013)); (**c**) cantilever (Source: Boisen and Thundat (2009)); (**d**) nanoparticles (Source: Sam et al. (2018)); (**e**) Nanowire (Source: Liu et al. (2017)); (**f**) nanopores (Source: Shi et al. (2016)); (**g**) nanorods (Source: Chen et al. (2013)) are used in the development of nanobiosensors

scattered from UV to NIR (Madani et al. 2013). Quantum dots are multifunctional and have fluorescent properties that are superior to conventional organic dyes (includes broad and narrow emission spectra). These also include photostability materials against photobleaching to tolerate changes in the pH of biological electrolytes (Chan and Nie 1998). The size and composition will determine whether the quantum dots are excited chemically in UV or NIR light. The structure of quantum dots contains the inorganic core, inorganic shell, and aqueous organic coating. The inorganic core contains elements that decide the wavelength of emitted light for clinical work to ascertain the toxicity and stability of the elements. Studies determine the toxicity, cell death, and DNA damaging properties (Ghaderi et al. 2011). Thus, quantum dots coupled with specific biomarkers can detect cancer cells.

10.2.2 Carbon Nanotubes

Carbon Nanotubes (CNT) are ultralightweight with the high surface area made of carbon atoms, which are chemical and thermal stability, and have high loading capacity (de La Zerda and Gambhir 2007). CNTs are single-walled or multi-walled with a single cylindrical carbon layer structure (0.2–2 nm diameter) (Madani et al. 2011). The binding of biological macromolecules to nanomaterial influences nanostructures due to minor surface perturbations and the novel electron transport properties help in electronic conductance offering real-time detection. Thus, detection of specific oncomarkers with sensitivity is possible with CNT (Madani et al. 2011).

10.2.3 Nanopores and Nanorods

Nanopores and nanorods have the potential for detecting drugs and biomolecules, and the nanopores are made up of protein or synthetic material. Molecules are passed through micropores and are subjected to the electric field. The concentration and nature of molecules can be detected by sensing individual molecules. Protein nanopores detect the immobilized DNA and also detect the methylation of DNA by epigenetic analysis (Clarke et al. 2009). Nanorods like gold nanorods are other ultrasensitive sensors potentially used for the detection of cancer biomarkers (Choi et al. 2010). Thus, nanopores and nanorods have a potential role in the detection of cancer biomarkers.

10.2.4 Nanowires and Cantilevers

Nanowires and cantilevers have a potential role in the detection of cancer biomarkers. Nanowires like silicon nanowires and gold conduction polymer nanowires detect lung cancer biomarkers (Hu et al. 2011). Cantilevers are potential biosensors that detect biomarkers based on nanometer-scale bends. Microcantilever assay helps in the detection of PSA, based on bend deflection caused by antigen-antibody binding due to the shift in frequency. Based on these shifts molecular interactions and target concentration can be determined (Hu et al. 2011). Thus, nanowires and cantilevers can be used for the detection of cancer biomarkers.

10.2.5 Nanoparticles

Nanoparticles like silica, gold, and magnetic nanoparticles are widely used in oncomarkers detection. Gold nanoparticles are potential imaging labels and agents which are used in the diagnosis or detection of microscopic tumors based on optical properties, resistance to photobleaching, and infrared light region absorption (Israelsen et al. 2015). Silica and magnetic nanoparticles are with thermophysical properties which are suitable for specific cancer marker diagnosis. These thermophysical properties help to absorb thermal energy without increasing temperature during phase change. These changes restrict the metal components inside the nanoparticles encapsulated with silica. In preclinical studies of colon cancer, the accumulation of silica-polythene glycol nanoparticles in tissue with two peptides specific to metastasis provided a powerful platform for the detection of micrometastases (Soster et al. 2012). Thus, nanoparticles can be used for the detection of cancer biomarkers.

The biomarkers used for the identification of various cancers like breast, colorectal, pancreatic, liver, prostate, lung, kidney, melanoma (Table 10.1) and gastrointestinal cancer (Table 10.2) are listed below.

Thus, nanomaterials like quantum dots, carbon nanotubes, nanopores, nanorods, nanowires, cantilevers, and nanoparticles can be employed for the development of nanobiosensors; and biomarkers listed above can be coupled with nanobiosensors for early detection of cancer especially gastrointestinal cancer.

10.3 Nanobiosensors

Literature reports the use of nanobiosensors like nanoparticles-based sensors, nanotube-based sensors, nanowire-based sensors, and ion channel-based sensors for detection of cancer.

10.3.1 Nanoparticles-Based Sensors

Nanoparticles-based sensors are classified as acoustic wave biosensors, magnetic biosensors, and electrochemical biosensors.

10.3.1.1 Acoustic Wave Biosensors

Acoustic wave biosensors are devices that detect a change of mass density, viscoelastic, elastic dielectric, or electric properties of materials made of chemically interactive with piezoelectric materials. Some commonly used transducers for this

Table 10.1 List of biomarkers used in the identification of cancer and types of nanomaterials employed for the development of nanobiosensors

S. no.	Biomarkers	Types of cancer	Types of nanomaterials	References
1	HER2/Neu	Breast cancer	Quantum dots	Dasilva et al. (2012)
2	CEA	Breast cancer	gold nanoparticles, quantum dots	Dasilva et al. (2012)
3	CEA	Colorectal cancer	Gold nanoparticles, quantum dots	Dasilva et al. (2012)
4	Peptide fragments	Colorectal cancer	Porous silicon nanoparticles	Li et al. (2014)
5	Matrix metalloproteinase	Colorectal cancer	Liposome (DPPC, MSPC, DPSE-PEG	Schuerle et al. (2016)
6	Mesothelin	Pancreatic cancer	Nanoparticle or theranostic nanoparticle drug carriers	Zhu et al. (2017)
7	Urokinase plasminogen activator	Pancreatic cancer	Nanoparticle or theranostic nanoparticle drug carriers	Zhu et al. (2017)
8	IGF-1R	Pancreatic cancer	Fluorescent imaging	Park et al. (2016)
9	EGFR	Pancreatic cancer	Single-chain anti-EGFR antibody-conjugated nanoparticles	Yang et al. (2009)
10	Plectin-1	Pancreatic cancer	Magnetofluorescent nanoparticles	Kelly et al. (2008)
11	Mucin-1	Pancreatic cancer	Superparamagnetic iron oxide nanoparticles	Moore et al. (2004)
12	Zinc transporter 4	Pancreatic cancer	Nanoparticle or theranostic nanoparticle drug carriers	Zhu et al. (2017)
13	α-fetoprotein	Cancer	AU nanoparticles	Kavosi et al. (2014)
14	PDGF-BB	Cancer	Au-PDMS	Zhu et al. (2016)
15	Volatile organic compound	Various cancer	Breath array nanosensors	Peng et al. (2010)
16	Sialic acids	Breast, liver cancer	Au nanoparticles	Zhang et al. (2016)

(continued)

Table 10.1 (continued)

S. no.	Biomarkers	Types of cancer	Types of nanomaterials	References
17	Protein biomarkers (EGFR, HER2, CD44, CD24)	Breast cancer	Au nanoparticles	Wang et al. (2016)
18	EVOM	Breast cancer	Fe3O4,SiO2 C18	Qiao et al. (2015)
19	Prostate-specific antigen	Prostate cancer	Au nanoparticles	Garcia-Cortes et al. (2016)
20	fPSA, cPSA	Prostate cancer	Au nanoparticles	Yoo and Yeo (2016)
21	Human IgG	Prostate cancer	Gold nanoparticles	Zheng et al. (2015)
22	CYFRA21-1, PSA	Lung, prostate cancer	Silicon nanowire	Lu et al. (2015)
23	Diglyceride, octadecanamide	Kidney cancer	Au nanoparticles	Nizioł et al. (2016)
24	EGFR, Her-2/Neu uPAR	Prostate, lung, breast, colorectal, and pancreas cancer	Silica and magnetic nanoparticles	Fruscella et al. (2016)
25	PMSA, AFP CA-125, CA-19,9 EpCAM, DNA-methyl	Prostate, pancreas, breast, lung cancer	Quantum dots	Fruscella et al. (2016)
26	FR, AFP	Prostate, liver, pancreas, breast, lung cancer	Carbon nanotubes	Fruscella et al. (2016)
27	L-6 CEA, ER, VEGF EpCAM, CK-7 IL-10, OPN CA-125 mRNA	Prostate, breast, lung cervical cancer	Nanowires	Fruscella et al. (2016)
28	SA, AFP, CEA, BRCA1 DNA/RNA	Prostate, breast, liver melanoma cancer	Cantilevers	Fruscella et al. (2016)
29	Hepsin, α-hemolysin	Prostate	Nanopores	Fruscella et al. (2016)
30	SA IL-10, VEGF	Prostate, lung cancer	Nanorods	Fruscella et al. (2016)

are bulk acoustic wave and surface acoustic wave transducers. Surface acoustic wave biosensors are inexpensive, flexible to point-of-care and real-time diagnostics. In an analysis of the sample, the devices incorporated in airway tubing capture molecules

Table 10.2 List of biomarkers used for identification of gastrointestinal cancer and types of nanomaterials employed in the development of nanobiosensors

S. no.	Biomarkers	Types of nanomaterials	References
1	CEA	Gold nanoparticles, quantum dots	Yoo and Yeo (2016)
2	MGC-803	Au-Ag nanoparticles	Zheng et al. (2015)
3	2-butanone	Au-Ag nanoparticles	Lu et al. (2015)
4	C-reactive proteins	Silicon nanowire arrays	Nizioł et al. (2016)
5	RCAS	Nanopores	Fruscella et al. (2016)
6	CA10-9	Nanopores	Fruscella et al. (2016)

in breath condensate. This approach is a very promising technology for both academic and industrial applications (Länge et al. 2008). Thus, acoustic wave biosensors can be employed for early detection of cancer.

10.3.1.2 Magnetic Biosensors

Magnetic biosensors are specialized magnetic nanoparticles mostly ferrite-based materials. The magnetic compounds have several analytic applications like the screening of iron coupled transition metals with different properties via the conjunction of magnetic nanoparticles. Magnetic biosensors are widely used in biomedical applications (Richardson et al. 2001) and some special devices like SQUID (superconducting quantum interference devices) are used for the detection of biological targets rapidly. The superparamagnetic nature of magnetic nanoparticles is used for screening specific antigens in a mixture of antibodies bounded to magnetic nanoparticles (Chemla et al. 2000). Thus, magnetic biosensors can be employed for early detection of cancer.

10.3.1.3 Electrochemical-Based Biosensors

The principle of this biosensor is that the reaction between the analyte and immobilized molecule consumes or produces ions that impact the electrical properties of solution which can be measured based on the electrochemical signal. The amount of analyte present in a given sample can also be measured quantitatively (Clark and Lyons 1962). This biosensor is the most successful commercialized sensors for multiple analytes which include pathogens and toxins. The other advantage of these biosensors is a high sensitivity, low cost, low power requirements, with miniaturization, turbidity, and color (Zhang et al. 2011). Thus, electrochemical-based biosensors can be employed for early detection of cancer.

10.3.2 Carbon Nanotube-Based Sensors

Carbon nanotube-based sensors are one-dimensional nanomaterials with unique structure-dependent electronic and mechanical properties. Novel electron transport properties are strongly influenced by the binding of macromolecules. These kinds of sensors offer real-time, sensitive, and label-free bioelectronics detection (Desai et al. 1999). Thus, carbon nanotube-based biosensors can be employed for early detection of cancer.

10.3.3 Nanowire-Based Sensors

Nanowire biosensors have major sensing components made up of nanowire-coated biomolecules called bionanowire, i.e., a fibril-like nanostructure. The surface properties are easily modified with chemical or biological recognition units when macromolecules bind to the surface of nanowires, and nanomaterials which transduce these changes in the conductance of nanowire. Nanowires are very efficient in excitation and transportation of elcctrons. These two are critical factors for the integration and function of nanodevices. Nanowires are very efficient for the development of enzyme, protein-based sensors based on their unique properties (Cui et al. 2001). Nanowires are also sensitive enough for the detection of an analyte. Thus, nanowire-based biosensors can be employed for early detection of cancer.

10.3.4 Ion Channel-Based Sensors

Ion channel-based biosensors play a prominent role in selectively and regulating the ions flow by controlling biochemical activities. The ion-based sensors mimic the biological sensory function and can be used for biological receptors including antibodies. This technology is very simple, flexible, and very sensitive (i.e., detection of a target in the picomolar concentration of proteins). It is helpful in cell typing, detection of large proteins, antibodies, compounds, and drugs (Cornell et al. 1997). Thus, ion channel-based biosensors can be employed for early detection of cancer.

Thus, nanobiosensors like nanoparticles-based sensors (acoustic wave biosensors, magnetic biosensors, and electrochemical biosensors), nanotube-based sensors, nanowire-based sensors, and ion channel-based sensors can be employed for detection of cancer especially gastrointestinal cancer.

10.4 Immobilization Strategies

The development of the first biosensor with nanotechnology had experienced tremendous growth in complexity, application, and utilization of devices. This increased the capability of detecting specific molecules with precision. Immobilization strategies in conjunction with biomolecules and nanomaterials developed various types of biosensors for real-time monitoring. The problem with the immobilization technique is the maintenance of biomolecule conformation and activity. The nonspecific molecule undergoes degradation which affects the function of a biosensor. There are different methods for immobilizing biomolecules onto nanostructure, and this is classified into three categories. The first category allowed the biomolecules to non-covalently bind with nanoparticles. The nanoparticles are with chemisorbed monolayer having a hydrophobic surface. In the next step, these nanoparticles are precipitated in water with tensidic micelles like phospholipids and sodium dodecyl sulfate. Finally, the molecules bind covalently to functional groups on the outer sphere of micelles. This category is advantageous as the whole process is easy to perform, as well also the interaction between nanostructure and biomolecule is based on hydrophobic interactions (Dubertret et al. 2002). The second category uses a linker to chemiabsorb biomolecules onto nanoparticles. The biomolecules with thiols groups directly help in chemisorption, where bifunctional molecules are coupled to these chemisorbed molecules that are similar to the first category approach (Bruchez et al. 1998). The third category allows the biomolecules to bind covalently to nanoparticles, these particles are derived with a cross-linked surface shell and with binding sites of functional groups. The biomolecule that binds to these surface shells is very stable due to covalent bonds. This category is very efficient for the long-term stability of the conjugate (Taton et al. 2001). The disadvantage of this method is that it is very expensive and difficult to immobilize. The major problem is that biomolecule becomes colloid when attached to nanoparticles, and there is a tendency for coagulation within biological data.

10.5 Parameters Indicating Performance of Nanobiosensors

The development of nanobiosensors requires optimization of individual elements and must be optimized to meet selectivity (response of the target biomolecules) which differentiates the specific and nonspecific reactions. Thus, nanobiosensors are suitable for detecting target molecules in lower concentrations. The nonspecific molecules may reduce the binding locations and create a false-positive signal. To overcome this problem (nonspecific binding reactions), blocking agents like bovine serum albumin can be used. The dose–response curve also called a calibration curve is used to detect the range of concentrations using nanobiosensor. Reproducibility, dynamic range, and negligible changes in concentrations of biomolecules are the parameters of nanobiosensors indicating the performance of nanobiosensors (Baszkin and Norde 1999).

10.6 Integration of Nanotechnology in Biochips

The integration of nanotechnology for the diagnosis will succeed in the analysis of biological and chemical information at a low cost. Some of the known nanotechnology-based biochips are protein nanobiochips, nanofluidic arrays, etc. Nanofluidic sensors are very promising in isolation and analysis of DNA and protein. This could lead to the development of new detection methods for chronic disease. This also plays a prominent role in drug development, personalized medicine, and other broad applications (Miekisch et al. 2004).

10.7 Applications of Nanobiosensors

Nanobiosensors have multiple and versatile applications like a diagnosis of diseases, environment monitoring, estimation of toxicity, and detection of carcinogens (Fig. 10.2).

10.7.1 Biomedical and Diagnostics Applications

Nanobiosensors had wide applications in the field of biomedical science and diagnosis of diseases. Nanobiosensors play a major role in the diagnosis of highly critical health problems that affect mankind. Diagnosis of some diseases like cancer is difficult, time-consuming, and highly expensive and biosensors improve the diagnosis of diseases (Gao et al. 2004). Biosensors are used in the detection of serum antigens, metabolic disorders, carcinogens, causative agents, diabetes, disorders, and many more. In the clinical point of view, nanobiosensors help in detecting the glucose of diabetic patients, bacterial infections of the urinary tract, HIV infections, and diagnosis of cancer. Serum analysis detects blood-related diseases and incorporation of the chip in sensing materials helps in the analysis of multiple diseases at a single go. Thus, nanobiosensors are useful in the diagnosis of disease.

10.7.2 Environmental Applications

Nanobiosensors have enormous applications in monitoring of environment and the changes in the environment can be detected for every second. Pollutants, toxic

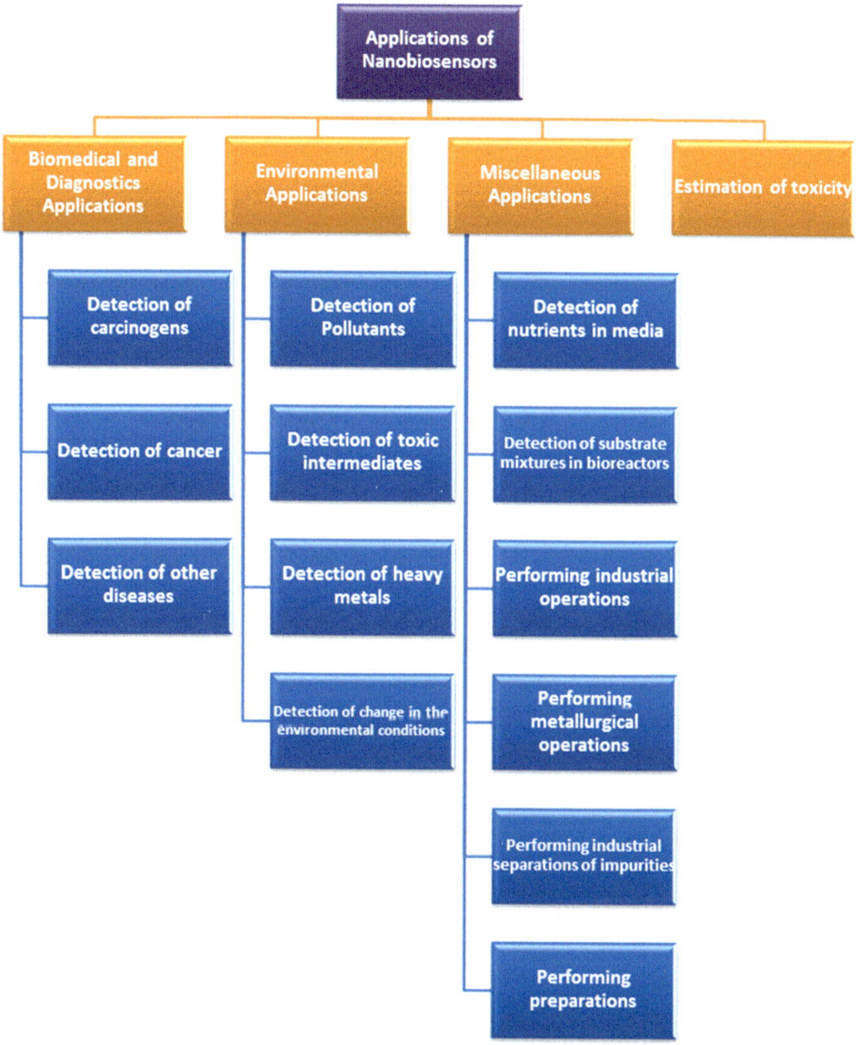

Fig. 10.2 Applications of nanobiosensors in the fields of biomedical and diagnostics, environment monitoring, estimation of toxicity, and miscellaneous applications

intermediates, and heavy metals change the environmental conditions. The nanomaterials-based devices like cantilever sensors are good sensing tools which can be used to find the damaging material present in the environment. Some studies on harmful intermediates that disrupt the hormonal system in living organisms are reported. Hamster ovary cell lines coupled with fluorescent materials reported harmful agents that affect the aqueous environment (Kim et al. 2002). Thus, nanobiosensors are very efficient and beneficial for monitoring of the environment.

10.7.3 Miscellaneous Applications

Nanobiosensors have many other applications like detection of nutrients in media and substrate mixtures in bioreactors; performing industrial operations, metallurgical operations, and industrial separations of impurities; and preparations using sensors. The above processes can also be regulated using nanobiosensors. Innovations based on the development of microbiological and biochemical assays are very reliable and easy to handle (Malik et al. 2013). Thus, nanobiosensors have a role in miscellaneous applications.

10.8 Nanobiosensors for Early Detection of Cancer

Nanobiosensors play a major role in the early detection of cancer. Carbon nanotubes are employed for the electrochemical and biological sensors based on their good electrochemical properties. The modified carbon nanotubes electrodes are used to immobilize biomolecules to minimize fouling effects. Recent studies showed evidence that carbon nanotubes promote electron transport reactions. CNTs also have properties like great strength, small size, large surface area, high conductivity, etc. Nanoparticles are suitable candidates when combined with paper-based sensing. Electroimmuno assay on 3D microfluidic paper-based device and multiplex electrochemiluminescence device a battery with microfluidic paper play a big role in the detection of cancer biomarkers (Zang et al. 2012). The paper surface can be modified with carbon paste, wax, carbon nanotubes, chitosan, and glutaraldehyde with the help of screen printing technique and wax. These devices are used for the detection of cancer biomarkers like CEA, and carcinoma antigen 199,125, etc. CEA of 0.05–50.0 ng mL and CA of 0.001–75.0 U mL (Wang et al. 2012) were detected in clinical samples using wax printing and screen printing. The volatile biomarkers distinguish MGS 803 gastric cancer cells and mucous cells based on Au-Ag nanomaterials. These volatile biomarkers detection is an early warning for gastric cancer (Zhang et al. 2014). The most effective environmentally friendly and promising platform is a breath sensor to screen early gastric cancer. The chloroplast is used as stabilizers and reducers for biosynthesizing Au-Ag alloy nanoscale. X-ray diffraction, UV-visible spectroscopy, etc. are used for the analysis to identify biomolecules in chloroplast for the Au-Ag alloy. The proteins in chloroplast bind to Au-Ag alloy through the amine group by using carbon nanotubes. The Au-Ag alloy is dispersed in multi-walled carbon nanotubes to form nanosensing film to exhibit the electrocatalytic activity of 2-butanone. This study provides a potential platform to screen biomarkers for stomach cancer (Zhang et al. 2012).

10.9 Nanobiosensors for Early Detection of Gastrointestinal Cancer

Gastrointestinal cancer is one of the leading cancers in the world and miR-106a is overexpressed in gastrointestinal cancer malignancies. Electrochemical nanobiosensor is used in the detection of miR-106a with the help of gold nanocomposite tag and double-specific probe methodology (Daneshpour et al. 2016). The hybridization of target miR and electrode modification is confirmed by cyclic voltammetry or differential pulse voltammetry or electrochemical impedance spectroscopy. These were used to evaluate and record the reduction peak of miR-106a. The concentration of target ranges from 1×10^{-3} pM to 1×10^{3} pM (Daneshpour et al. 2016). The nanosensor in sample investigation had a great performance, high specificity, and remarkable selectivity. Thus, nanosensors provide the most promising applications like the detection of gastrointestinal cancer.

10.10 Conclusion and Future Directions

The current challenge for effective treatment of gastrointestinal cancer is the detection of cancer at an early stage. Detection of cancer at an early stage requires biomarkers expressing at an early stage and also promising technologies like nanotechnology. Nanotechnology is a powerful and promising technology for early detection of gastrointestinal cancer. The nanomaterials used for biosensors are quantum dots, carbon nanotubes, nanopores, nanorods, nanowires, cantilevers, nanoparticles, and nanomembranes. These nanomaterials are used in the development of nanobiosensors which can be used in the detection of cancer and gastrointestinal cancer. Nanobiosensors role in early detection of cancer via carbon nanotube for CEA biomarker and nanobiosensors role in for early detection of gastrointestinal cancer via electrochemical nanobiosensor for biomarkers miRNA 106A were reported. The other different applications of nanobiosensors are biomedical and diagnostics applications, environmental applications, and miscellaneous applications. The different nanobiosensors are nanoparticles-based sensors (acoustic wave biosensors, magnetic biosensors, electrochemical biosensors), nanotube-based sensors, nanowire-based sensors, and ion channel-based sensors. The immobilization of biomolecules onto nanomaterials develops nanobiosensors for the detection of the analyte. The different strategies used for immobilization of biomolecules onto nanomaterials are covalent, noncovalent, and linker with covalent. The different parameters like selectivity, reproducibility, dynamic range, and negligible changes in concentrations of biomolecules indicate the performance of nanobiosensors. There is a need for real-time measurements and integration of nanotechnology or nanobiosensors in biochips will help in real-time measurements. This also leads to the development of devices at a low cost. The world has already witnessed the development of wearable biosensors like glucometers for health care monitoring which help in real-time measurements of glucose in diabetic patients. Future can also witness the integration of nanotechnology to develop nanobiosensors and transformation of these biosensors into wearable biosensors for health care monitoring, i.e., early detection of gastrointestinal cancer based on biomarkers which are expressed at an early stage of cancer. Thus, it can be concluded that nanobiosensors have a potential role in the early detection of gastrointestinal cancer.

Acknowledgments ND and NNRR are grateful to GITAM (Deemed-to-be-University) for providing necessary facilities to carry out the research work and for extending constant support.

Conflict of Interest The authors declare that there is no potential conflict of interest.

References

Arnold MA, Meyerhoff ME (1988) Recent advances in the development and analytical applications of biosensing probes. Crit Rev Anal Chem 20(3):149–196

Baszkin A, Norde W (1999) Physical chemistry of biological interfaces. CRC Press, Boca Raton

Boisen A, Thundat T (2009) Design and fabrication of cantilever array biosensors. Mater Today 12 (9):32–38

Bondy M (2009) Cancer epidemiology and prevention. JAMA 301(10):1070–1076

Bruchez M, Moronne M, Gin P, Weiss S, Alivisatos AP (1998) Semiconductor nanocrystals as fluorescent biological labels. Science 281(5385):2013–2016

Chan WC, Nie S (1998) Quantum dot bioconjugates for ultrasensitive nonisotopic detection. Science 281(5385):2016–2018

Chemla YR, Grossman HL, Poon Y, McDermott R, Stevens R, Alper MD, Clarke J (2000) Ultrasensitive magnetic biosensor for homogeneous immunoassay. Proc Natl Acad Sci U S A 97(26):14268–14272

Chen H, Shao L, Li Q, Wang J (2013) Gold nanorods and their plasmonic properties. Chem Soc Rev 42:2679–2724

Choi YE, Kwak JW, Park JW (2010) Nanotechnology for early cancer detection. Sensors 10 (1):428–455

Clark LC Jr, Lyons C (1962) Electrode systems for continuous monitoring in cardiovascular surgery. Ann N Y Acad Sci 102(1):29–45

Clarke J, Wu HC, Jayasinghe L, Patel A, Reid S, Bayley H (2009) Continuous base identification for single-molecule nanopore DNA sequencing. Nat Nanotechnol 4(4):265–270

Cornell BA, Braach-Maksvytis VLB, King LG, Osman PDJ, Raguse B, Wieczorek L, Pace RJ (1997) A biosensor that uses ion-channel switches. Nature 387(6633):580–583

Cui Y, Wei Q, Park H, Lieber CM (2001) Nanowire nanosensors for highly sensitive and selective detection of biological and chemical species. Science 293(5533):1289–1292

Cui DX, Zhang L, Yan XJ, Zhang LX, Xu JR, Guo YH, Jin GQ, Gomez G, Li D, Zhao JR, Han FC (2005) A microarray-based gastric carcinoma prewarning system. World J Gastroenterol 11 (9):1273–1282

Daneshpour M, Omidfar K, Ghanbarian H (2016) A novel electrochemical nanobiosensor for the ultrasensitive and specific detection of femtomolar level gastric cancer biomarker miRNA-106a. Beilstein J Nanotechnol 7(2016):2023–2036

Dasilva N, Díez P, Matarraz S, González-González M, Paradinas S, Orfao A, Fuentes M (2012) Biomarker discovery by novel sensors based on nanoproteomics approaches. Sensors 12 (2):2284–2308 de La Zerda A, Gambhir SS (2007) Drug delivery: keeping tabs on nanocarriers. Nat Nanotechnol 2 (12):745–746

Demchenko AP, Dekaliuk MO (2013) Novel fluorescent carbonic nanomaterials for sensing and imaging. Methods Appl Fluoresc 1(4):042001. https://doi.org/10.1088/2050-6120/1/4/042001

Desai T, Hansford D, Ferrari M (1999) Characterization of micromachined silicon membranes for immunoisolation and bioseparation applications. J Memb Sci 159:221–231

Di Giusto DA, Wlassoff WA, Gooding JJ, Messerle BA, King GC (2005) Proximity extension of circular DNA aptamers with real-time protein detection. Nucleic Acids Res 33(6):e64–e64

Dubertret B, Skourides P, Norris DJ, Noireaux V, Brivanlou AH, Libchaber A (2002) In vivo imaging of quantum dots encapsulated in phospholipid micelles. Science 298(5599):1759–1762

Ferrari M (2005) Cancer nanotechnology: opportunities and challenges. Nat Rev Cancer 5 (3):161–171

Fruscella M, Ponzetto A, Crema A, Carloni G (2016) The extraordinary progress in very early cancer diagnosis and personalized therapy: the role of oncomarkers and nanotechnology. J Nanotechnol 2016:18. https://doi.org/10.1155/2016/3020361

Gao X, Cui Y, Levenson RM, Chung LW, Nie S (2004) In vivo cancer targeting and imaging with semiconductor quantum dots. Nat Biotechnol 22(8):969–976

Garcia-Cortes M, Encinar JR, Costa-Fernandez JM, Sanz-Medel A (2016) Highly sensitive nanoparticle-based immunoassays with elemental detection: application to prostate-specific antigen quantification. Biosens Bioelectron 85:128–134

Ghaderi S, Ramesh B, Seifalian AM (2011) Fluorescence nanoparticles "quantum dots" as drug delivery system and their toxicity: a review. J Drug Target 19(7):475–486

Hanahan D, Weinberg RA (2000) The hallmarks of cancer. Cell 100(1):57–70

Higgins MJ, Ettinger DS (2009) Chemotherapy for lung cancer: the state of the art in 2009. Expert Rev Anticancer Ther 9(10):1365–1378

Hu Y, Fine DH, Tasciotti E, Bouamrani A, Ferrari M (2011) Nanodevices in diagnostics. Wiley Interdiscip Rev Nanomed Nanobiotechnol 3(1):11–32

Israelsen ND, Hanson C, Vargis E (2015) Nanoparticle properties and synthesis effects on surface-enhanced Raman scattering enhancement factor: an introduction. Sci World J 2015:124582. https://doi.org/10.1155/2015/124582

Kavosi B, Hallaj R, Teymourian H, Salimi A (2014) Au nanoparticles/PAMAM dendrimer functionalized wired ethyleneamine–viologen as highly efficient interface for ultra-sensitive α-fetoprotein electrochemical immunosensor. Biosens Bioelectron 59:389–396

Kelly KA, Bardeesy N, Anbazhagan R, Gurumurthy S, Berger J, Alencar H, DePinho RA, Mahmood U, Weissleder R (2008) Targeted nanoparticles for imaging incipient pancreatic ductal adenocarcinoma. PLoS Med 5(4):e85. https://doi.org/10.1371/journal.pmed.0050085

Kim EJ, Lee Y, Lee JE, Gu MB (2002) Application of recombinant fluorescent mammalian cells as a toxicity biosensor. Water Sci Technol 46(3):51–56

Länge K, Rapp BE, Rapp M (2008) Surface acoustic wave biosensors: a review. Anal Bioanal Chem 391(5):1509–1519

Li X, Tan J, Yu J, Feng J, Pan A, Zheng S, Wu J (2014) Use of a porous silicon–gold plasmonic nanostructure to enhance serum peptide signals in MALDI-TOF analysis. Anal Chim Acta 849:27–35

Liu R, Chen R, Elthakeb AT, Lee SH, Hinckley S, Khraiche ML, Scott J, Pre D, Hwang Y, Tanaka A, Ro YG, Matsushita AK, Dai X, Soci C, Biesmans S, James A, Nogan J, Jungjohann KL, Pete DV, Webb DB, Zou Y, Bang AG, Dayeh SA (2017) High density individually addressable nanowire arrays record intracellular activity from primary rodent and human stem cell derived neurons. Nano Lett 17:2757–2764

Lu N, Gao A, Dai P, Mao H, Zuo X, Fan C, Wang Y, Li T (2015) Ultrasensitive detection of dual cancer biomarkers with integrated cmos-compatible nanowire arrays. Anal Chem 87 (22):11203–11208

Madani SY, Naderi N, Dissanayake O, Tan A, Seifalian AM (2011) A new era of cancer treatment: carbon nanotubes as drug delivery tools. Int J Nanomedicine 6:2963–2979

Madani SY, Shabani F, Dwek MV, Seifalian AM (2013) Conjugation of quantum dots on carbon nanotubes for medical diagnosis and treatment. Int J Nanomedicine 8:941–950
Malik P, Katyal V, Malik V, Asatkar A, Inwati G, Mukherjee TK (2013) Nanobiosensors: concepts and variations. ISRN Nanomater 2013:1–9
Miekisch W, Schubert JK, Noeldge-Schomburg GF (2004) Diagnostic potential of breath analysis—focus on volatile organic compounds. Clin Chim Acta 347(1–2):25–39
Moore A, Medarova Z, Potthast A, Dai G (2004) In vivo targeting of underglycosylated MUC-1 tumor antigen using a multimodal imaging probe. Cancer Res 64(5):1821–1827
Nizioł J, Ossoliński K, Ossoliński T, Ossolińska A, Bonifay V, Sekuła J, Dobrowolski Z, Sunner J, Beech I, Ruman T (2016) Surface-transfer mass spectrometry imaging of renal tissue on gold nanoparticle enhanced target. Anal Chem 88(14):7365–7371
Park JY, Lee JY, Zhang Y, Hoffman RM, Bouvet M (2016) Targeting the insulin growth factor-1 receptor with fluorescent antibodies enables high resolution imaging of human pancreatic cancer in orthotopic mouse models. Oncotarget 7(14):18262–18268
Peng G, Hakim M, Broza YY, Billan S, Abdah-Bortnyak R, Kuten A, Tisch U, Haick H (2010) Detection of lung, breast, colorectal, and prostate cancers from exhaled breath using a single array of nanosensors. Br J Cancer 103(4):542–551
Phillips M, Cataneo RN, Saunders C, Hope P, Schmitt P, Wai J (2010) Volatile biomarkers in the breath of women with breast cancer. J Breath Res 4(2):026003. https://doi.org/10.1088/1752-7155/4/2/026003
Prasad S (2014) Nanobiosensors: the future for diagnosis of disease. Nanobiosens Dis Diagn 3:1–10
Qiao Z, Perestrelo R, Reyes-Gallardo EM, Lucena R, Cárdenas S, Rodrigues J, Câmara JS (2015) Octadecyl functionalized core–shell magnetic silica nanoparticle as a powerful nanocomposite sorbent to extract urinary volatile organic metabolites. J Chromatogr A 1393:18–25
Reddy RK (2007) Nanomonitors: electrical immunoassay for clinical diagnostics implementation of a microfabricated biosensor for the detection of proteins. Master of Science Thesis, Portland State University, Portland
Richardson J, Hawkins P, Luxton R (2001) The use of coated paramagnetic particles as a physical label in a magnetoimmunoassay. Biosens Bioelectron 16(9–12):989–993
Salvati E, Stellacci F, Krol S (2015) Nanosensors for early cancer detection and for therapeutic drug monitoring. Nanomedicine 10(23):3495–3512
Sam C, Jacob K, Spencer L (2018) Creating nanoparticle shapes and properties. https://www.materialstoday.com/biomaterials/articles/s1369702109702494/
Schuerle S, Dudani JS, Christiansen MG, Anikeeva P, Bhatia SN (2016) Magnetically actuated protease sensors for in vivo tumor profiling. Nano Lett 16(10):6303–6310
Shi J, Hou J, Fang Y (2016) Recent advances in nanopore-based nucleic acid analysis and sequencing. Microchim Acta 183(3):925–939
Soster M, Juris R, Bonacchi S, Genovese D, Montalti M, Rampazzo E, Zaccheroni N, Garagnani P, Bussolino F, Prodi L, Marchiò S (2012) Targeted dual-color silica nanoparticles provide univocal identification of micrometastases in preclinical models of colorectal cancer. Int J Nanomedicine 7:4797–4807
Taton TA, Lu G, Mirkin CA (2001) Two-color labeling of oligonucleotide arrays via size-selective scattering of nanoparticle probes. J Am Chem Soc 123(21):5164–5165
Wang S, Ge L, Zhang Y, Song X, Li N, Ge S, Yu J (2012) Battery-triggered microfluidic paper-based multiplex electrochemiluminescence immunodevice based on potential-resolution strategy. Lab Chip 12(21):4489–4498
Wang YW, Doerksen JD, Kang S, Walsh D, Yang Q, Hong D, Liu JT (2016) Multiplexed molecular imaging of fresh tissue surfaces enabled by convection-enhanced topical staining with SERS-coded nanoparticles. Small 12(40):5612–5621
Yang L, Mao H, Wang YA, Cao Z, Peng X, Wang X, Duan H, Ni C, Yuan Q, Adams G, Smith MQ (2009) Single chain epidermal growth factor receptor antibody conjugated nanoparticles for in vivo tumor targeting and imaging. Small 5(2):235–243

Yoo M, Yeo WS (2016) Determining the ratio of two types of prostate specific antigens with biochips and gold nanoparticles for accurate prostate cancer diagnosis. Anal Sci 32 (10):1117–1121

Zang D, Ge L, Yan M, Song X, Yu J (2012) Electrochemical immunoassay on a 3D microfluidic paper-based device. Chem Commun 48:4683–4685

Zhang X, Ju H, Wang J (eds) (2011) Electrochemical sensors, biosensors and their biomedical applications. Academic, Cambridge

Zhang Y, Gao G, Qian Q, Cui D (2012) Chloroplasts-mediated biosynthesis of nanoscale Au-Ag alloy for 2-butanone assay based on electrochemical sensor. Nanoscale Res Lett 7(1):475. https://doi.org/10.1186/1556-276X-7-475

Zhang Y, Gao G, Liu H, Fu H, Fan J, Wang K, Chen Y, Li B, Zhang C, Zhi X, He L (2014) Identification of volatile biomarkers of gastric cancer cells and ultrasensitive electrochemical detection based on sensing interface of Au-Ag alloy coated MWCNTs. Theranostics 4 (2):154–162

Zhang X, Chen B, He M, Zhang Y, Peng L, Hu B (2016) Boronic acid recognition based-gold nanoparticle-labeling strategy for the assay of sialic acid expression on cancer cell surface by inductively coupled plasma mass spectrometry. Analyst 141(4):1286–1293

Zheng T, Pierre-Pierre N, Yan X, Huo Q, Almodovar AJ, Valerio F, Rivera-Ramirez I, Griffith E, Decker DD, Chen S, Zhu N (2015) Gold nanoparticle-enabled blood test for early stage cancer detection and risk assessment. ACS Appl Mater Interfaces 7(12):6819–6827

Zhu D, Yang RX, Tang YP, Li W, Miao ZY, Hu Y, Chen J, Yu S, Wang J, Xu CY (2016) Robust nanoplasmonic substrates for aptamer macroarrays with single-step detection of PDGF-BB. Biosens Bioelectron 85:429–436

Zhu L, Staley C, Kooby D, El-Rays B, Mao H, Yang L (2017) Current status of biomarker and targeted nanoparticle development: the precision oncology approach for pancreatic cancer therapy. Cancer Lett 388:139–148

Part III
Computational Methods for Identification Gastrointestinal Cancer Biomarkers and Early Diagnosis of GI Cancer

Chapter 11
Genetic Marker Identification for the Detection of Early-Onset Gastric Cancer Through Genome-Wide Association Studies

Manoj Kumar Gupta, Jinka Rajeswari, Pamuru Ramachandra Reddy, Koppula Satish Kumar, K. V. Chamundeswaramma, and Ramakrishna Vadde

Abstract The complete human genome sequence published by Celera and Human Genome Project in 2001 has provided us with in-depth knowledge about both location and structure of genes; however, they do not provide any information about the genetic diversity between and within human populations. International associations such as the 1000 genomes project, Simons Genome Diversity Project and International HapMap project employed high-throughput sequencing technologies to explore the genetic diversity among various human population across the world. All these studies suggested that every human endures 250–350 loss-of-function mutations on average as well as they are heterozygous for about 60–120 variants, which are associated with genetic disorders. Information about this genetic diversity among human enable us to carry out genome-wide association studies (GWAS) and recognize genes and its respective variants related with any traits of interest or diseases. Till date, more than 1600 GWAS studies have been reported on ~300 traits and diseases. Gastric cancer is a solid tumor with complex genetic and

M. K. Gupta · R. Vadde (✉)
Department of Biotechnology & Bioinformatics, Yogi Vemana University, Kadapa, Andhra Pradesh, India
e-mail: vrkbtbi@yogivemanauniversity.ac.in

J. Rajeswari
Department of Biochemistry, Acharya Nagarjuna University, Guntur, Andhra Pradesh, India

P. R. Reddy
Department of Biochemistry, Yogi Vemana University, Kadapa, Andhra Pradesh, India

K. S. Kumar
Department of Biotechnology, Adikavi Nannaya University, Rajamahendravaram, Andhra Pradesh, India

K. V. Chamundeswaramma
Department of Zoology, Government College for Men, Anantapur, Andhra Pradesh, India

P. Veera Bramhachari, N. R. R. Neelapu (eds.), *Recent Advancements in Biomarkers and Early Detection of Gastrointestinal Cancers*, Diagnostics and Therapeutic Advances in GI Malignancies, https://doi.org/10.1007/978-981-15-4431-6_11

environmental interactions. Major gastric cancers are adenocarcinomas. Germline disparities of DNA sequence between different ethnic populations make one population more disposed to gastric cancer as compared to others. In this article, we reviewed all genes and its polymorphisms related with early-onset gastric carcinoma. In near future, these identified genetic polymorphisms can be utilized as biomarker in the detection and prevention of gastric cancer.

Keywords Early-onset gastric cancer · GWAS · Biomarker

11.1 Introduction

Cancer is clinically characterized via uncontrolled growth of abnormal cells (Yi et al. 2016). As cancer can develop from abnormal growth of any kinds of cells in the body, there are numerous distinct types of cancer (Cooper 2000). However, lung, breast, colorectal, prostate, skin (non-melanoma), and gastric cancer are the common most cancers found in human body (Bray et al. 2018). All forms of cancer can be divided into three groups, namely carcinomas (arise in the skin/tissues lining the internal organs), sarcomas (develop in the muscle, bone, fat, cartilage, or other connective tissues), and leukemia (develops in the blood and bone marrow) (Cooper 2000). Out of these three, only in carcinomas and sarcomas, abnormal growth of cell form tumor.

One of the key features that differentiates cancer cells from normal cell is their ability to spread throughout the body via invasion and metastasis. While invasion is the direct extension and penetration of cancer cells into its neighboring tissues, metastasis is the movement of cancer cells from its origin site to another site. During metastasis, cancer cells penetrate into blood or lymphatics vessels and while circulating through these systems invades normal tissues in the body. As this metastasis phenomenon occurs in an predictable and orderly manner, sometimes this whole phenomenon is called as the "metastatic cascade" (Lloyd et al. 2017). Metastases is the main cause of death from cancer (Bray et al. 2018).

Cancer is the second leading cause of death in the United States (Siegel et al. 2018). In 2017, 1,735,350 total new cases have been registered for all types of cancer, and 609,640 individuals have died due to various cancer. Out of 1,735,350 registered new cases, 26,240 are of gastric cancer (Siegel et al. 2018). Like other cancer, apart from genetic factors, non-genetic factors like alcohol consumption, smoking, physical inactivity, poor diet, and stress enhance the risk of being affected by gastric cancer (Guggenheim and Shah 2012). Recent population-based study reported that men are more associated with tobacco smoking and alcohol consumption, and are more prone to gastric cancer than women. Among men, East and West Asian are more prone to gastric cancer than Northern America and Northern Europe (Bray et al. 2018). Apart from these abiotic factors, biotic factors like *Helicobacter pylori* (*H. pylori*) and Epstein–Barr virus (EBV) also play a significant role in the etiology of gastric cancer. In 1994, *H. pylori* was categorized as a class I carcinogen by the World Health Organization (Kusters et al. 2006). EBV is a ubiquitous human herpes virus which is often associated with numerous lymphoid as well as epithelial malignancies, including Hodgkin's lymphoma, Burkitt's lymphoma, nasal NK/T cell lymphoma, a subset of gastric carcinomas, and nasopharyngeal carcinoma (Cho et al. 2016). *H. pylori* is a microaerophilic Gram-negative bacterium that colonizes the gastric mucosa of 50% of the human population (Ishaq and Nunn 2015). As gastric cancer is often diagnosed at an advanced stage, little progress is seen during the treatment of advanced or metastatic gastric cancer and median overall survival remains less than 1 year (Carcas 2014). Hence, there is an urgent need for future research to detect genetic changes during early-onset gastric cancer to improvize the outcomes of this killer disease.

Recent advancement of high-throughput sequencing technologies like microarray and genome-wide association study (GWAS) enable us to detect genes along with its alleles, which are responsible for causing various diseases. These genes along with their alleles serve as a biomarker in the prevention or curing of any diseases (Gupta et al. 2017a, b). Recent study reported that gastric cancer when detected at an early stage, it is generally curable and the five-year survival rate is generally >90%. Early-onset gastric cancer is generally confined to the mucosa or submucosa irrespective of the existence of lymph node metastasis. However, because of nonspecific symptoms and trouble in characterizing early gastric cancer from benign peptic ulcer or gastritis in the ambulatory setting, only <20% of gastric cancer are diagnosed at an early stage globally (Pirini et al. 2017). Hence, in the present review we will discuss in brief about various genes and its variant(s) which get differentially expressed during early onset of gastric cancer. In near future, these genes along with its variant(s) may serve as a biomarker in the prevention of gastric cancer in human.

11.2 Classification of Gastric Cancer

Gastric cancer can be categorized based on anatomic location, nature of occurrence, and pathology.

11.2.1 Based on Pathology

As per Lauren classification, gastric cancer can be subdivided into two groups, namely intestinal and diffuse types (Lauren 1965). Intestinal type related with *H. pylori*-associated chronic gastritis, atrophy, and intestinal metaplasia. Though diffuse type is also related with *H. pylori* infection, but it is not associated with atrophy and intestinal metaplasia. In diffuse type, atrophy and intestinal metaplasia are less differentiated, characterized via sheets of cells deprived of gland formation, with the irregular presence of signet ring cells and mucin, and are related with a poor diagnosis compared with the intestinal type (van der Woude et al. 2003). Incident of both diffuse and intestinal types of gastric cancer varies among population. Europeans are more prone to intestinal type, particularly diffuse type occurs mostly in younger patients (Sitarz et al. 2018).

11.2.2 Based on Anatomic Location

Cancer may develop at proximal stomach/cardia (cardia) or at mid or distal stomach (non-cardia) (Mukaisho et al. 2015). Colquhoun and team reported that incident of cardia and non-cardia gastric cancers is higher in central Asia and Eastern/South-eastern Asia, respectively (Colquhoun et al. 2015). Cigarette smoking (Cook et al. 2010) and low intake of fresh fruits and vegetables (Freedman et al. 2008) causes both cardia and non-cardia gastric cancer. Esophageal adenocarcinoma, obesity, Barrett's esophagus (a metaplastic condition which develops from gastroesophageal reflux disease), and gastroesophageal reflux disease are also responsible for causing cardia gastric cancer in some population (Hoyo et al. 2012). For cardia gastric cancer, male to female ratio is 3:1, and thus males are at higher risk for cardia gastric cancer than females (Colquhoun et al. 2015). Non-cardia gastric cancer is strongly associated with *H. pylori* infection (Plummer et al. 2015). Some studies also reported that low socioeconomic status is positively co-related with non-cardia gastric cancer (Guggenheim and Shah 2012).

11.2.3 Based on Occurrence

Gastric cancer can be either early onset or advanced. Early-onset gastric cancer occurs in individual before the age of 45 years and comprises of about 10% of gastric cancer. Early gastric cancer is more common in females (Derakhshan et al. 2009) and is generally diffuse type (Kokkola and Sipponen 2001). They often characterized as multifocal (Carneiro et al. 2004), infrequent loss of heterozygosity (Carvalho et al. 2004), lack of microsatellite instability (Carvalho et al. 2004), lack of intestinal metaplasia (Matley et al. 1988), loss of *TFF1* expression (Milne et al. 2006), and the presence of *RUNX3* (Carvalho et al. 2005). About 10% of early-onset gastric cancers have positive family history (Kokkola and Sipponen 2001). Though EBV and *H. pylori* involved in causing gastric cancer, their impact is less on early-onset gastric cancer (Milne and Offerhaus 2010). This type of cancer is mainly caused via genetic factors, for instance, mutation in *CDH1* germline encodes abnormal E-cadherin, resulting in hereditary diffuse gastric cancer (Carneiro et al. 2008).

Advanced gastric cancer is the most common type of gastric cancer and mainly develop in people over the age of 45 years. As they occur sporadically, they are often called "sporadic gastric cancers." They are mainly caused via environmental factors and host genetic factors (Skierucha et al. 2016). In 1975, Correa and team reported that sporadic gastric cancer is related with progressive premorbid histological modification in the gastric epithelium (Correa et al. 1975). The initial modification includes superficial gastritis wherein the gastric mucosa is penetrated with both acute and chronic inflammatory cells. This inflammatory mechanism consequently turns into more powerful within the acid secreting area of the stomach and reduces glands as well as acid-secreting parietal cell. Subsequent loss of gastric acid secretion causes colonization of the stomach via mixed bacterial flora (Correa et al. 1975). Later studies reported that *H. pylori* infection is the real cause of the superficial gastritis (Suriani et al. 2008). Individual with more virulent Cag A-positive strain of infection experiences high atrophic hypochlorhydria pattern of gastritis (Suriani et al. 2008). A diet with less fresh fruits and high in salt accelerate the progression to atrophic gastritis (Kusters et al. 2006). Though smoking is a general risk factor for gastric cancer, it does not enhance the rate of growth of atrophic gastritis or hypochlorhydria. As the men are more associated with smoking and more prone to advanced gastric cancer.

11.3 Genome-Wide Association Study and Cancer

Genome-wide association study investigates DNA sequence variations across whole genome to recognize genetic factors responsible for developing any trait or causing any diseases in a specific population. One of the classical example of early successes of GWAS was the recognition of the *Complement Factor H* gene as an important risk factor for age-associated muscular degeneration (Fisher et al. 2005; Haines et al. 2005). Until today, >1600 GWAS studies have been published on approximately 250 traits and disease, including skin color and body weight. These studies reported that every human individual bears 250–350 loss-of-function mutations on average and are heterozygous for 60–120 variants responsible for genetic disorders (Bush and Moore 2012).

Over the past few decades, more than 700 risk loci associated with various types of cancer have been discovered in different human population. Out of 700 risk loci, ~80%, ~15%, and <1% loci are responsible for causing cancer in East Asian, European, and African and Latin American populations (Sud et al. 2017). Till date, through GWAS identified well-validated several risk loci for different cancers, including prostate (Al Olama et al. 2014), breast (Michailidou et al. 2015), colorectal (Orlando et al. 2016), pancreatic (Wolpin et al. 2014), lung (McKay et al. 2017), gastric (Helgason et al. 2015), renal (Scelo et al. 2017), and bladder (Rothman et al. 2010). Risk loci have also been identified in malignant melanoma (Law et al. 2015), ovarian cancer (Kuchenbaecker et al. 2015), basal cell carcinoma (Chahal et al. 2016), glioma (Melin et al. 2017), testicular germcell tumor (Wang et al. 2017), Hodgkin lymphoma (Cozen et al. 2014), thyroid cancer (Gudmundsson et al. 2017), chronic lymphocytic leukemia (CLL) (Law et al. 2017), follicular lymphoma (Skibola et al. 2014), meningioma (Dobbins et al. 2011), multiple myeloma (Mitchell et al. 2016), and diffuse large B cell lymphoma (Cerhan et al. 2014). Furthermore, common risk alleles have also been recognized via GWAS for numerous pediatric solid cancers, including neuroblastoma (Diskin et al. 2009) and Wilms tumor (Turnbull et al. 2012).

Due to large sample size in almost all GWAS studies performed on breast and prostate cancer, a number of risk loci identified for breast and prostate cancer are greater than other cancers (Michailidou et al. 2015). Difference in the heritability of different cancers influences the identification of risk loci differently. For instance, chronic lymphocytic leukemia is highly heritable and has an eightfold familial relative risk (Goldin et al. 2004) while lung cancer is non-heritable (McKay et al. 2017). As chronic lymphocytic leukemia is heritable, GWAS of only 17,598 controls and 6200 patients identified 43 risk loci for chronic lymphocytic leukemia (Law et al. 2017). On the contrary, GWAS of 56,450 controls and 29,266 patients identified only 18 risk loci associated with all lung cancer subtypes, suggesting the significance of non-genetic risk factors in the pathophysiology of lung cancer (McKay et al. 2017).

11.4 Molecular Mechanism of Gastric Cancer

Gastric cancer is generally detected at advance stage and occur via both environmental factors, like *H. pylori* infection, and the genetic modifications (Grabsch and Tan 2013). While the infections via *H. pylori* are asymptomatic, they are often related with peptic ulcers, gastritis, and the most severe form of gastric cancer. *H. pylori* induces DNA damage and abnormal methylation patterns, which in turn affects the downstream cell signaling of the host. Apart from that, *H. pylori* also triggers epithelial-to-mesenchymal transition and increases the likelihood of pro-survival and proliferation signals which contribute to the cancer phenotype (Servetas et al. 2016). Though epithelial-to-mesenchymal transition is dangerous for the developmental process, it plays a significant physio-pathological role in epithelial tumorigenesis (Larue and Bellacosa 2005). During epithelial-to-mesenchymal transition, epithelial cell loses its normal properties, which are required for maintaining the healthy cellular barrier, and behaves like mesenchymal cells, for instance, with increased mobility, loss of cellular interactions, and loss of polarization (Larue and Bellacosa 2005). *H. pylori* infection is also reported to decrease the expression of E-cadherin (epithelial marker) and increase the expression of *SNAIL*, *TWIST*, *SLUG*, and vimentin (mesenchymal markers) causing epithelial-to-mesenchymal transition (Choi et al. 2015). Some other studies reported that suppression of miR-328, which is a negative regulator of CD44 expression, via H_2O_2 enhanced CD44 expression, which in turn enhances the expression of mesenchymal markers leading to gastric cancer (Ishimoto et al. 2015). Several other studies reported that when CD44, CagA, and c-Met form a complex, they enhance cellular proliferation and expression of mesenchymal markers, which in turn disrupt the normal functions of membrane-associated E-cadherin causing gastric cancer. Another studies reported that when RBP2 binds with p-SMAD3 on the promoter region of E-cadherin, it reduces E-cadherin expression, which in turn causes gastric cancer (Liang et al.

2015). Silencing of *TFF1* and *RPRM* gene and overexpression of HER2, CD44v6, and *MUC2* are also responsible for causing gastric cancer (Fig. 11.1).

11.5 Biomarker for Early-Onset Gastric Cancer

Though the underlying complete genetic mechanism is not clear till date, several studies reported that *CDH1* (Bacani et al. 2006), *CTNNB1* (Zhou et al. 2002), *CDX-2* (Fan et al. 2005), *HER2* (Moelans et al. 2011), CD44v6 (Milne and Offerhaus 2010), 5p15 (Du et al. 2013), *PRKAA1* (Jiang et al. 2018), and *Reprimo* (Bernal et al. 2008) may serve as biomarkers for detecting early-onset gastric cancer.

11.5.1 CDH1

The *CDH1* gene is located at 16q22.1 and is made up of 16 exons (Berx et al. 1995). *CDH1* encodes a glycoprotein, namely E-cadherin, which is mostly localized at adherens junctions of epithelial cells and is responsible for modulating homophilic calcium-dependent cell-adhesion (Shapiro et al. 1995). As E-cadherin is a tumor suppressor protein, downregulation of E-cadherin causes several cancers (Berx and van Roy 2009). Mutation is one of the most important events involved in silencing tumor suppressor genes, somatic mutations of *CDH1* have been reported to cause several cancer, including lobular breast cancer (Huiping et al. 1999), ovarian cancer (Risinger et al. 1994), sporadic diffuse gastric cancer (Becker and Höfler 1995), and colorectal cancer (Efstathiou et al. 1999). However, studies of familial gastric cancer, lacking cancers in other organs, proposed that germline mutation is the main cause for inducing inherited gastric cancer (Carneiro et al. 2008). For the first time, germline mutation of *CDH1* was identified in the DNA obtained from lymphocytes of two patients with gastric cancer and four obligate carriers in New Zealand. The investigation of exon 2 to exon 16 of *CDH1* gene utilizing the single-stranded conformational polymorphism technique suggested that a band shift in exon 7. Later, direct sequencing of exon 7 suggested germline mutation at 1008 base position of *CDH1* from G to T induces early-onset gastric cancer (Guilford et al. 1998). Later, several studies identified various mutations associated with early onset of gastric cancer, in *CDH1* gene. Oliveira and team identified 1901 C>T variant in exon 12 of Portuguese (Oliveira et al. 2004). Zhang and team identified 2253 C>T variant in exon 14 of Chinese (Zhang et al. 2006). Nasri and team identified 163 +37235G>A variant in intron 2 of Italian (Nasri et al. 2008). Kim and team identified 1003 C>T variants in exon 7 of Italian (Kim et al. 2013). All these mutations either truncate E-cadherin protein or cause abnormal modifications of the E-cadherin's calcium binding sites, or increase its proteolytic degradation, which in turn disrupt the normal functions of E-cadherin (Liu and Chu 2014). Deactivation of E-cadherin reduces cell–cell adhesion and initiates abnormal modification of

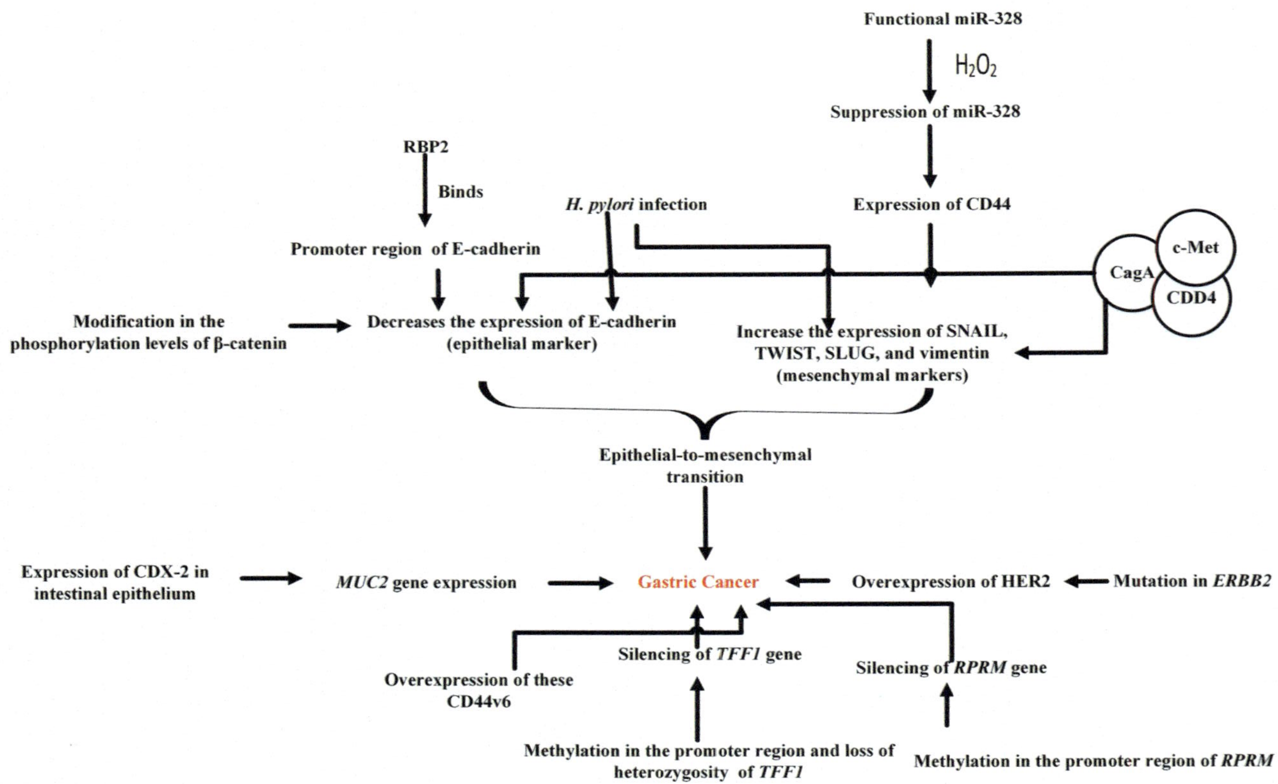

Fig. 11.1 Molecular mechanism involved in gastric cancer. Epithelial-to-mesenchymal transition due to decrease in expression of E-cadherin (epithelial marker) and increase in expression of *SNAIL, TWIST, SLUG*, and vimentin (mesenchymal markers) causes gastric cancer. Silencing of *TFF1* and *RPRM* gene and overexpression of HER2, CD44v6, and *MUC2* are also responsible for causing gastric cancer

E-cadherin-associated signaling pathways associated with cell proliferation, EMT process, and inflammation. These abnormal modification initiates early gastric cancer development (Liu and Chu 2014).

11.5.2 *CTNNB1*

CTNNB1 gene encodes β-catenin (Sineva and Pospelov 2014), which is a dual functional protein and is expressed mainly in cytoplasm, membrane, and nucleus of epithelial cells. Nuclear β-catenin plays a significant role in the canonical Wnt/wingless pathway. It aggregates in the nucleus and along with other genes of LEF-1/TCF family activates important genes responsible for differentiation and cellular proliferation (Behrens 1999). On the other hand, the membranous and cytosolic β-catenin together with E-cadherin and the actin cytoskeleton work as a chief constituent of cell–cell adhesive junctions (Retterspitz et al. 2010). Modification in the phosphorylation levels of β-catenin modifies the function of E-cadherin, which in turn initiates early gastric cancer development (Chiurillo 2015).

11.5.3 *Caudal Type Homeobox-2* (CDX-2)

CDX-2 is a caudal-related homeobox transcription factor. In adults, its expression is restricted nearby the intestinal epithelium and is mainly involved in the development as well as the maintenance of intestinal mucosa (Suh et al. 1994). *CDX-2* mRNA is generally overexpressed in the caecum and colon with reduced expression levels in other intestine regions. *CDX-2* is not expressed in the normal gastric mucosa (Mizoshita et al. 2001). Earlier Mesquita and team reported that as the promoter region of mucin 2 (*MUC2*) contains CDX putative binding region, expression of *CDX-2* in intestinal epithelium may trigger the *MUC2* gene expression in gastric cell, which in turn causes both gastric and colon cancer (Mesquita et al. 2003). Another comparative study between *CDX-2* expression and dysplasia in cancer tissues suggested that as, in comparison to diffuse gastric cancer, expression level of *CDX-2* expression is higher in intestinal metaplasia tissues (Mizoshita et al. 2003). *CDX-2* expression may serve as a biomarker for the early onset of gastric cancer. Likewise, many other studies also suggested *CDX-2* as a positive prognostic factor for an early onset of gastric cancer (Halder et al. 2018).

11.5.4 *HER2*

HER2 is one of the four tyrosine receptor kinases of EGFR family (EGFR or HER1, HER2, HER3, and HER4) encoded via proto-oncogene *ERBB2* situated on chromosome 17q21. It plays a significant role in the survival as well as proliferation of cell (Normanno et al. 2005; Durães et al. 2014). For transmitting signal, HER2 must heterodimerize with other HER family members, especially with EGFR (Ou 2012). Early studies reported that mutation in *ERBB2* gene causes overexpression of HER2 during early stage of carcinogenesis (Fassan et al. 2012), which in turn causes survival, growth, and proliferation of cancer cells via the PI3K-AKT as well as the MAPK pathways (Gravalos and Jimeno 2008; Gallardo et al. 2012). Though it is well established that overexpression of HER2 receptor is biomarker in both breast and gastric cancer (Baniak et al. 2016), level of HER2 overexpression depends on the location and histology of the cancer (Dragovich et al. 2006; Gravalos and Jimeno 2008). In comparison to distal gastric location, HER2 overexpression is more frequent at gastroesophageal junction tumors, and it is generally associated with the intestinal type adenocarcinomas (Kim et al. 2007; Kunz et al. 2012).

11.5.5 TFF1

The trefoil peptides, namely trefoil factor 1 (*TFF1*), *TFF2*, and *TFF3*, are a group of highly conserved small proteins present mainly in mucous granules in mucus-secreting cells. Trefoil peptides are mainly expressed and secreted via epithelial cells lining mucous membranes (Thim and May 2005). During mucosal injury, *TFF1* overexpression maintains the integrity of the mucosa and stomach ontogenesis (Soutto et al. 2011). Some studies reported *TFF1* as a candidate tumor suppressor gene (Calnan et al. 1999). Other studies reported that methylation of the *TFF1* promoter region and loss of heterozygosity (LOH) silence *TFF1* gene, which in turn cause gastric cancer (Park et al. 2000). Silencing of *TFF1* is also reported to be initiated via chromatin remodeling related with histone modifications, for instance, H3 deacetylation and H3K9 methylation at the *TFF1* promoter, as observed in *N*-methyl-*N*-nitrosourea-induced gastric carcinogenesis mouse model (Tomita et al. 2011). *TFF1* silencing within gastric epithelial cells is also reported to occur via CCAAT/enhancer binding protein-β (Sankpal et al. 2005) and cofactor of *BRCA1* (McChesney et al. 2006). However, the range of histological lesions and the molecular mechanism that mediated by the loss of *TFF1* in gastric tumorigenesis remain unclear till date.

11.5.6 Reprimo

The Reprimo gene family is a new single-exon intron less gene family (Amigo et al. 2018). Reprimo is mainly related with developmental patterning of the gastrointestinal tract, brain, and blood vessels (Wichmann et al. 2016; Figueroa et al. 2017). Reprimo functions as tumor suppression gene. During DNA damage, p53 induced up-regulation of Reprimo arrest cell cycle at the G2/M checkpoint (Ohki et al. 2000). However, epigenetic silencing of *RPRM* via DNA methylation of its promoter region initiates early stages of human cancer (Amigo et al. 2018).

11.5.7 Lack of Microsatellite Instability

Microsatellite DNA are randomly distributed short and repetitive DNA sequences in the human genome (Leung et al. 2000). Inactivation of mismatch repair genes, along with *hMLH1* and *hMSH2*, which prevents repair of replication errors, for instance, deletion or insertion of bases in microsatellite regions is known as microsatellite instability. Microsatellite instability is also caused through epigenetic promoter methylation (Fang et al. 2012). Earlier studies reported that microsatellite instability usually occur in hereditary nonpolyposis colorectal cancer and advanced gastric cancer (Aaltonen et al. 1993; Milne and Offerhaus 2010) and is absent during early-onset gastric cancer (Milne and Offerhaus 2010). In most of the gastric cancer, microsatellite instability develops due to hypermethylation of the MLH1 promoter (Fleisher et al. 1999).

11.5.8 CD44v6

CD44 is a polymorphic membrane glycoprotein localized mainly in human cell associated with T-cell activation, cellular matrix adhesion, and lymphocyte residing in specific lymph node tissue (Carvalho et al. 2006). Earlier studies reported that CD44 is mainly involved in triggering the cytoskeletal rearrangements as well as morphological modification required for active migration of tumor cells in the extracellular matrix, which in turn is highly required for invasiveness and metastasis (Marhaba and Zöller 2004). Expression of these CD44v6 is associated with up-regulation of anti-apoptotic genes, which in turn causes various cancer. Earlier studies reported that overexpression of CD44v6 is associated with tumor progression, invasion, and metastatic behavior in early-onset gastric carcinomas (Carvalho et al. 2006).

11.5.9 5p15

Earlier GWAS reported that genetic variants situated at chromosome 5p15 are risk factors for various cancer (Du et al. 2013). First two GWAS reported that this locus, comprising of *TERT* gene, is involved in lung cancer (Wang et al. 2008). Though rs401681-C allele at 5p15 has a protective effect on cutaneous melanoma, it is a risk factor for lung, basal cell, prostate, cervical, and urinary bladder cancer (Rafnar et al. 2009). Later, locus of 5p15 was also associated with risk of glioma (Skibola et al. 2014), pancreatic (Petersen et al. 2010), and breast cancers (Haiman et al. 2011). Du and team reported variant rs10052016 at 5p15 is significantly associated with early-onset gastric cancer risk in Chinese population (Du et al. 2013).

11.5.10 *PRKAA1*

AMP-activated protein kinase catalytic subunit alpha-1 gene (*PRKAA1*) is located on chromosome 5p13.1 and encodes 5′-AMP-activated protein kinase (AMPK) (Yamazaki et al. 2007). AMPK is an energy sensor and plays a significant role in the biosynthesis of macromolecules and cellular metabolism (Hardie et al. 1997). AMPK is activated via various molecular mechanisms, including G1 phase arrest in the cell cycle, inhibition of protein, and fatty acid synthesis (Jiang et al. 2018). AMPK activation inhibits accumulation of lipid in the body, enhances the oxidation of fatty acids, and reduces the synthesis of cholesterol as well as fatty acids (Hardie 2005). AMPK activation also suppresses cell proliferation in cancerous and nonmalignant cells (Jiang et al. 2018). In two independent studies, shi and team and Helgason and team reported that variant rs13361707 T>C of *PRKAA1* is a risk factor for gastric cancer in both European and Asian population (Chen et al. 2018). As variant rs13361707 alters mRNA expression *PRKAA1,* Jiang and team hypothesize that *PRKAA1* may serve as biomarker for the early onset of gastric cancer (Jiang et al. 2018).

11.5.11 microRNA in Gastric Juice

microRNAs are small RNA segments responsible for regulating the expression of numerous genes at the post-transcriptional level and function as potential antioncogenes (tumor-suppressor miRs) as well as oncogenes (oncogenic miRs or oncomiRs). Earlier studies reported that few microRNAs, like *miR-421, miR-129,* and *miR-133a*, present in the tissue (mucosa) or serum gastric cancer patients may serve as biomarkers in the detection of early-onset gastric cancer. *miR-421* is a well-known oncomiR associated with up-regulation of tumor-associated nuclear receptors (Li et al. 2018). Recently, Yang and team reported that *miR-421* enhances the proliferation, invasion, and metastasis of gastric cancer cells via targeting Claudin 11 (*CLDN11*) (Yang et al. 2017). In another study, Wu and team reported that *miR-421* overexpression enhances cell growth as well as suppressed apoptosis in gastric cancer cells (Wu et al. 2014). *miR-421* overexpression is also reported to cause nasopharyngeal carcinoma through downregulating forkhead box protein O4 (*FOXO4*) (Chen et al. 2013) and biliary tract cancer via downregulating farnesoid

X receptor (*FXR*) (Zhong et al. 2012). *miR-129* functions as a tumor suppressor via modulating G_1/S phase transition and apoptosis (Yu et al. 2013). *miR-133a* is mainly localized in muscle (hence called myomiR) and promotes muscle growth, modulating initial differentiation of myogenic stem cells into myoblasts. *miR-133a* also functions as a tumor-suppressor gene (Wang et al. 2014). Less expression of both *miR-129* and *miR-133a* may cause gastric cancer (Virgilio et al. 2018).

11.6 Conclusion

Recent advancement of high-throughput sequencing technologies like microarray and genome-wide association study enable us to detect genes along with its alleles, which are responsible for causing various diseases. These genes along with their alleles may serve as a biomarker in the prevention or curing of any diseases (Gupta et al. 2017a, b). Though several studies have reported numerous novel biomarkers to predict various stages of gastric cancer, only few biomarkers, like carcinoembryonic antigen and carbohydrate antigen 19-9 are in clinical use. As these biomarkers are associated with advance stage of gastric cancer, little progress is seen during the treatment of advanced or metastatic gastric cancer and median overall survival remains less than 1 year (Carcas 2014). Hence, it is highly required to develop biomarkers that are highly specific, noninvasive, conventional, capable of early detection as well as leading to treatment choice. Recently, genome studies reported that expression level of *CDH1* (Suriano et al. 2003; Bacani et al. 2006), *CTNNB1* (Zhou et al. 2002), *CDX-2* (Seno et al. 2002; Mizoshita et al. 2003; Fan et al. 2005), *HER2* (Moelans et al. 2011), CD44v6 (Carvalho et al. 2006), 5p15 (Du et al. 2013), *PRKAA1* (Jiang et al. 2018), and *Reprimo* (Bernal et al. 2008) in gastric region can be utilized as a predictive biomarker for the early onset of gastric cancer, which in near future can be utilized in curing/controlling gastric cancer.

11.7 Future Prospective

As gastric cancer is generally identified at advance stage, little progress is seen during treatment and thus median overall survival remains less than 1 year. Biomarker discussed in this chapter will enable medical practitioners or researcher to identify early-onset gastric cancer, which may increase the overall survival of the patients. Protein biomarkers like E-cadherin and CD44v6 may also be utilized as a target protein for developing drug against gastric cancer.

Conflict of Interests Authors declare no conflicts of interest.

References

Aaltonen LA, Peltomäki P, Leach FS et al (1993) Clues to the pathogenesis of familial colorectal cancer. Science 260:812–816

Al Olama AA, Kote-Jarai Z, Berndt SI et al (2014) A meta-analysis of 87,040 individuals identifies 23 new susceptibility loci for prostate cancer. Nat Genet 46:1103–1109. https://doi.org/10.1038/ng.3094

Amigo JD, Opazo JC, Jorquera R et al (2018) The Reprimo gene family: a novel gene lineage in gastric cancer with tumor suppressive properties. Int J Mol Sci 19:1862. https://doi.org/10.3390/ijms19071862

Bacani JT, Soares M, Zwingerman R et al (2006) CDH1/E-cadherin germline mutations in early-onset gastric cancer. J Med Genet 43:867–872. https://doi.org/10.1136/jmg.2006.043133

Baniak N, Senger J-L, Ahmed S et al (2016) Gastric biomarkers: a global review. World J Surg Oncol 14:212. https://doi.org/10.1186/s12957-016-0969-3

Becker K-F, Höfler H (1995) Frequent somatic allelic inactivation of the E-cadherin gene in gastric carcinomas. JNCI J Natl Cancer Inst 87:1082–1084. https://doi.org/10.1093/jnci/87.14.1082

Behrens J (1999) Cadherins and catenins: role in signal transduction and tumor progression. Cancer Metastasis Rev 18:15–30. https://doi.org/10.1023/A:1006200102166

Bernal C, Aguayo F, Villarroel C et al (2008) Reprimo as a potential biomarker for early detection in gastric cancer. Clin Cancer Res 14:6264–6269. https://doi.org/10.1158/1078-0432.CCR-07-4522

Berx G, van Roy F (2009) Involvement of members of the cadherin superfamily in cancer. Cold Spring Harb Perspect Biol 1:a003129. https://doi.org/10.1101/cshperspect.a003129

Berx G, Cleton-Jansen AM, Nollet F et al (1995) E-cadherin is a tumour/invasion suppressor gene mutated in human lobular breast cancers. EMBO J 14:6107–6115

Bray F, Ferlay J, Soerjomataram I et al (2018) Global cancer statistics 2018: GLOBOCAN estimates of incidence and mortality worldwide for 36 cancers in 185 countries. CA Cancer J Clin 68:394–424. https://doi.org/10.3322/caac.21492

Bush WS, Moore JH (2012) Chapter 11: genome-wide association studies. PLoS Comput Biol 8: e1002822. https://doi.org/10.1371/journal.pcbi.1002822

Calnan DP, Westley BR, May FE et al (1999) The trefoil peptide TFF1 inhibits the growth of the human gastric adenocarcinoma cell line AGS. J Pathol 188:312–317. https://doi.org/10.1002/(SICI)1096-9896(199907)188:3<312::AID-PATH360>3.0.CO;2-P

Carcas LP (2014) Gastric cancer review. J Carcinog 13:14. https://doi.org/10.4103/1477-3163.146506

Carneiro F, Huntsman DG, Smyrk TC et al (2004) Model of the early development of diffuse gastric cancer in E-cadherin mutation carriers and its implications for patient screening. J Pathol 203:681–687. https://doi.org/10.1002/path.1564

Carneiro F, Oliveira C, Suriano G, Seruca R (2008) Molecular pathology of familial gastric cancer, with an emphasis on hereditary diffuse gastric cancer. J Clin Pathol 61:25–30. https://doi.org/10.1136/jcp.2006.043679

Carvalho R, Milne ANA, van Rees BP et al (2004) Early-onset gastric carcinomas display molecular characteristics distinct from gastric carcinomas occurring at a later age. J Pathol 204:75–83. https://doi.org/10.1002/path.1602

Carvalho R, Milne ANA, Polak M et al (2005) Exclusion of RUNX3 as a tumour-suppressor gene in early-onset gastric carcinomas. Oncogene 24:8252–8258. https://doi.org/10.1038/sj.onc.1208963

Carvalho R, Milne ANA, Polak M et al (2006) A novel region of amplification at 11p12-13 in gastric cancer, revealed by representational difference analysis, is associated with overexpression of CD44v6, especially in early-onset gastric carcinomas. Genes Chromosomes Cancer 45:967–975. https://doi.org/10.1002/gcc.20360
Cerhan JR, Berndt SI, Vijai J et al (2014) Genome-wide association study identifies multiple susceptibility loci for diffuse large B cell lymphoma. Nat Genet 46:1233–1238. https://doi.org/10.1038/ng.3105
Chahal HS, Wu W, Ransohoff KJ et al (2016) Genome-wide association study identifies 14 novel risk alleles associated with basal cell carcinoma. Nat Commun 7:12510. https://doi.org/10.1038/ncomms12510
Chen L, Tang Y, Wang J et al (2013) miR-421 induces cell proliferation and apoptosis resistance in human nasopharyngeal carcinoma via downregulation of FOXO4. Biochem Biophys Res Commun 435:745–750. https://doi.org/10.1016/j.bbrc.2013.05.056
Chen M, Jiang B, He B et al (2018) Genetic variations in PRKAA1 predict the risk and progression of gastric cancer. BMC Cancer 18:923. https://doi.org/10.1186/s12885-018-4818-3
Chiurillo MA (2015) Role of the Wnt/β-catenin pathway in gastric cancer: an in-depth literature review. World J Exp Med 5:84–102. https://doi.org/10.5493/wjem.v5.i2.84
Cho J, Kang M-S, Kim K-M (2016) Epstein-Barr virus-associated gastric carcinoma and specific features of the accompanying immune response. J Gastric Cancer 16:1–7. https://doi.org/10.5230/jgc.2016.16.1.1
Choi YJ, Kim N, Chang H et al (2015) Helicobacter pylori-induced epithelial-mesenchymal transition, a potential role of gastric cancer initiation and an emergence of stem cells. Carcinogenesis 36:553–563. https://doi.org/10.1093/carcin/bgv022
Colquhoun A, Arnold M, Ferlay J et al (2015) Global patterns of cardia and non-cardia gastric cancer incidence in 2012. Gut 64:1881–1888. https://doi.org/10.1136/gutjnl-2014-308915
Cook MB, Kamangar F, Whiteman DC et al (2010) Cigarette smoking and adenocarcinomas of the esophagus and esophagogastric junction: a pooled analysis from the international BEACON consortium. JNCI J Natl Cancer Inst 102:1344–1353. https://doi.org/10.1093/jnci/djq289
Cooper GM (2000) The development and causes of cancer. In: The cell: a molecular approach, 2nd edn. Sinauer Associates, Sunderland
Correa P, Haenszel W, Cuello C et al (1975) A model for gastric cancer epidemiology. Lancet 2:58–60
Cozen W, Timofeeva MN, Li D et al (2014) A meta-analysis of Hodgkin lymphoma reveals 19p13.3 TCF3 as a novel susceptibility locus. Nat Commun 5:3856. https://doi.org/10.1038/ncomms4856
Derakhshan MH, Liptrot S, Paul J et al (2009) Oesophageal and gastric intestinal-type adenocarcinomas show the same male predominance due to a 17 year delayed development in females. Gut 58:16–23. https://doi.org/10.1136/gut.2008.161331
Diskin SJ, Hou C, Glessner JT et al (2009) Copy number variation at 1q21.1 associated with neuroblastoma. Nature 459:987–991. https://doi.org/10.1038/nature08035
Dobbins SE, Broderick P, Melin B et al (2011) Common variation at 10p12.31 near MLLT10 influences meningioma risk. Nat Genet 43:825–827. https://doi.org/10.1038/ng.879
Dragovich T, McCoy S, Fenoglio-Preiser CM et al (2006) Phase II trial of Erlotinib in gastroesophageal junction and gastric adenocarcinomas: SWOG 0127. J Clin Oncol 24:4922–4927. https://doi.org/10.1200/JCO.2006.07.1316
Du J, Xu Y, Dai J et al (2013) Genetic variants at 5p15 are associated with risk and early onset of gastric cancer in Chinese populations. Carcinogenesis 34:2539–2542. https://doi.org/10.1093/carcin/bgt259
Durães C, Almeida GM, Seruca R et al (2014) Biomarkers for gastric cancer: prognostic, predictive or targets of therapy? Virchows Arch 464:367–378. https://doi.org/10.1007/s00428-013-1533-y
Efstathiou JA, Liu D, Wheeler JMD et al (1999) Mutated epithelial cadherin is associated with increased tumorigenicity and loss of adhesion and of responsiveness to the motogenic trefoil factor 2 in colon carcinoma cells. Proc Natl Acad Sci U S A 96:2316–2321. https://doi.org/10.1073/pnas.96.5.2316

Fan Z, Li J, Dong B, Huang X (2005) Expression of Cdx2 and hepatocyte antigen in gastric carcinoma: correlation with histologic type and implications for prognosis. Clin Cancer Res 11:6162–6170. https://doi.org/10.1158/1078-0432.CCR-05-0278

Fang W-L, Chang S-C, Lan Y-T et al (2012) Microsatellite instability is associated with a better prognosis for gastric cancer patients after curative surgery. World J Surg 36:2131–2138. https://doi.org/10.1007/s00268-012-1652-7

Fassan M, Mastracci L, Grillo F et al (2012) Early HER2 dysregulation in gastric and oesophageal carcinogenesis. Histopathology 61:769–776. https://doi.org/10.1111/j.1365-2559.2012.04272.x

Figueroa RJ, Carrasco-Avino G, Wichmann IA et al (2017) Reprimo tissue-specific expression pattern is conserved between zebrafish and human. PLoS One 12:e0178274. https://doi.org/10.1371/journal.pone.0178274

Fisher SA, Abecasis GR, Yashar BM et al (2005) Meta-analysis of genome scans of age-related macular degeneration. Hum Mol Genet 14:2257–2264

Fleisher AS, Esteller M, Wang S et al (1999) Hypermethylation of the hMLH1 gene promoter in human gastric cancers with microsatellite instability. Cancer Res 59:1090–1095

Freedman ND, Subar AF, Hollenbeck AR et al (2008) Fruit and vegetable intake and gastric cancer risk in a large United States prospective cohort study. Cancer Causes Control 19:459–467. https://doi.org/10.1007/s10552-007-9107-4

Gallardo A, Lerma E, Escuin D et al (2012) Increased signalling of EGFR and IGF1R, and deregulation of PTEN/PI3K/Akt pathway are related with trastuzumab resistance in HER2 breast carcinomas. Br J Cancer 106:1367–1373. https://doi.org/10.1038/bjc.2012.85

Goldin LR, Pfeiffer RM, Li X, Hemminki K (2004) Familial risk of lymphoproliferative tumors in families of patients with chronic lymphocytic leukemia: results from the Swedish family-cancer database. Blood 104:1850–1854. https://doi.org/10.1182/blood-2004-01-0341

Grabsch HI, Tan P (2013) Gastric cancer pathology and underlying molecular mechanisms. Dig Surg 30:150–158. https://doi.org/10.1159/000350876

Gravalos C, Jimeno A (2008) HER2 in gastric cancer: a new prognostic factor and a novel therapeutic target. Ann Oncol 19:1523–1529. https://doi.org/10.1093/annonc/mdn169

Gudmundsson J, Thorleifsson G, Sigurdsson JK et al (2017) A genome-wide association study yields five novel thyroid cancer risk loci. Nat Commun 8:14517. https://doi.org/10.1038/ncomms14517

Guggenheim DE, Shah MA (2012) Gastric cancer epidemiology and risk factors. J Surg Oncol 107:230–236. https://doi.org/10.1002/jso.23262

Guilford P, Hopkins J, Harraway J et al (1998) E-cadherin germline mutations in familial gastric cancer. Nature 392:402–405. https://doi.org/10.1038/32918

Gupta MK, Behara SK, Vadde R (2017a) In silico analysis of differential gene expressions in biliary stricture and hepatic carcinoma. Gene 597:49–58

Gupta MK, Behera SK, Dehury B, Mahapatra N (2017b) Identification and characterization of differentially expressed genes from human microglial cell samples infected with Japanese encephalitis virus. J Vector Borne Dis 54:131–138

Haiman CA, Chen GK, Vachon CM et al (2011) A common variant at the TERT-CLPTM1L locus is associated with estrogen receptor-negative breast cancer. Nat Genet 43:1210–1214. https://doi.org/10.1038/ng.985

Haines JL, Hauser MA, Schmidt S et al (2005) Complement factor H variant increases the risk of age-related macular degeneration. Science 308:419–421

Halder A, Kundu M, Das RN et al (2018) CDX2 expression in gastric carcinoma: a clinicopathological study. Indian J Med Paediatr Oncol 39:52. https://doi.org/10.4103/ijmpo.ijmpo_49_17

Hardie DG (2005) New roles for the LKB1—>AMPK pathway. Curr Opin Cell Biol 17:167–173. https://doi.org/10.1016/j.ceb.2005.01.006

Hardie DG, Corton J, Ching YP et al (1997) Regulation of lipid metabolism by the AMP-activated protein kinase. Biochem Soc Trans 25:1229–1231

Helgason H, Rafnar T, Olafsdottir HS et al (2015) Loss-of-function variants in ATM confer risk of gastric cancer. Nat Genet 47:906–910. https://doi.org/10.1038/ng.3342

Hoyo C, Cook MB, Kamangar F et al (2012) Body mass index in relation to oesophageal and oesophagogastric junction adenocarcinomas: a pooled analysis from the international BEACON consortium. Int J Epidemiol 41:1706–1718. https://doi.org/10.1093/ije/dys176

Huiping C, Sigurgeirsdottir JR, Jonasson JG et al (1999) Chromosome alterations and E-cadherin gene mutations in human lobular breast cancer. Br J Cancer 81:1103–1110. https://doi.org/10.1038/sj.bjc.6690815

Ishaq S, Nunn L (2015) Helicobacter pylori and gastric cancer: a state of the art review. Gastroenterol Hepatol 8:S6–S14

Ishimoto T, Izumi D, Watanabe M et al (2015) Chronic inflammation with Helicobacter pylori infection is implicated in CD44 overexpression through miR-328 suppression in the gastric mucosa. J Gastroenterol 50:751–757. https://doi.org/10.1007/s00535-014-1019-y

Jiang Y, Li W, Lu J et al (2018) Association between PRKAA1 rs13361707 T>C polymorphism and gastric cancer risk: evidence based on a meta-analysis. Medicine (Baltimore) 97:e0302. https://doi.org/10.1097/MD.0000000000010302

Kim MA, Jung EJ, Lee HS et al (2007) Evaluation of HER-2 gene status in gastric carcinoma using immunohistochemistry, fluorescence in situ hybridization, and real-time quantitative polymerase chain reaction. Hum Pathol 38:1386–1393. https://doi.org/10.1016/j.humpath.2007.02.005

Kim S, Chung J-W, Jeong T-D et al (2013) Searching for E-cadherin gene mutations in early onset diffuse gastric cancer and hereditary diffuse gastric cancer in Korean patients. Fam Cancer 12:503–507. https://doi.org/10.1007/s10689-012-9595-6

Kokkola A, Sipponen P (2001) Gastric carcinoma in young adults. Hepatogastroenterology 48:1552–1555

Kuchenbaecker KB, Ramus SJ, Tyrer J et al (2015) Identification of six new susceptibility loci for invasive epithelial ovarian cancer. Nat Genet 47:164–171. https://doi.org/10.1038/ng.3185

Kunz PL, Mojtahed A, Fisher GA et al (2012) Her2 expression in gastric and gastroesophageal junction adenocarcinoma in a us population: clinicopathologic analysis with proposed approach to Her2 assessment. Appl Immunohistochem Mol Morphol 20:13–24. https://doi.org/10.1097/PAI.0b013e31821c821c

Kusters JG, van Vliet AHM, Kuipers EJ (2006) Pathogenesis of Helicobacter pylori infection. Clin Microbiol Rev 19:449–490. https://doi.org/10.1128/CMR.00054-05

Larue L, Bellacosa A (2005) Epithelial-mesenchymal transition in development and cancer: role of phosphatidylinositol 3′ kinase/AKT pathways. Oncogene 24:7443–7454. https://doi.org/10.1038/sj.onc.1209091

Lauren P (1965) The two histological main types of gastric carcinoma: diffuse and so-called intestinal-type carcinoma. An attempt at a HISTO-clinical classification. Acta Pathol Microbiol Scand 64:31–49

Law MH, Bishop DT, Lee JE et al (2015) Genome-wide meta-analysis identifies five new susceptibility loci for cutaneous malignant melanoma. Nat Genet 47:987–995. https://doi.org/10.1038/ng.3373

Law PJ, Berndt SI, Speedy HE et al (2017) Genome-wide association analysis implicates dysregulation of immunity genes in chronic lymphocytic leukaemia. Nat Commun 8:14175. https://doi.org/10.1038/ncomms14175

Leung WK, Kim JJ, Kim JG et al (2000) Microsatellite instability in gastric intestinal metaplasia in patients with and without gastric cancer. Am J Pathol 156:537–543. https://doi.org/10.1016/S0002-9440(10)64758-X

Li Y, Cui X, Li Y et al (2018) Upregulated expression of miR-421 is associated with poor prognosis in non-small-cell lung cancer. Cancer Manag Res 10:2627–2633. https://doi.org/10.2147/CMAR.S167432

Liang X, Zeng J, Wang L et al (2015) Histone demethylase RBP2 promotes malignant progression of gastric cancer through TGF-β1-(p-Smad3)-RBP2-E-cadherin-Smad3 feedback circuit. Oncotarget 6:17661–17674. https://doi.org/10.18632/oncotarget.3756

Liu X, Chu K-M (2014) E-cadherin and gastric cancer: cause, consequence, and applications. Biomed Res Int 2014:637308. https://www.hindawi.com/journals/bmri/2014/637308/. Accessed 22 Nov 2018

Lloyd MC, Gatenby RA, Brown JS (2017) Chapter 11: Ecology of the metastatic process. In: Ujvari B, Roche B, Thomas F (eds) Ecology and evolution of cancer. Academic, Cambridge, pp 153–165

Marhaba R, Zöller M (2004) CD44 in cancer progression: adhesion, migration and growth regulation. J Mol Histol 35:211–231. https://doi.org/10.1023/B:HIJO.0000032354.94213.69

Matley PJ, Dent DM, Madden MV, Price SK (1988) Gastric carcinoma in young adults. Ann Surg 208:593–596

McChesney PA, Aiyar SE, Lee O-J et al (2006) Cofactor of BRCA1: a novel transcription factor regulator in upper gastrointestinal adenocarcinomas. Cancer Res 66:1346–1353. https://doi.org/10.1158/0008-5472.CAN-05-3593

McKay JD, Hung RJ, Han Y et al (2017) Large-scale association analysis identifies new lung cancer susceptibility loci and heterogeneity in genetic susceptibility across histological subtypes. Nat Genet 49:1126–1132. https://doi.org/10.1038/ng.3892

Melin BS, Barnholtz-Sloan JS, Wrensch MR et al (2017) Genome-wide association study of glioma subtypes identifies specific differences in genetic susceptibility to glioblastoma and non-glioblastoma tumors. Nat Genet 49:789–794. https://doi.org/10.1038/ng.3823

Mesquita P, Jonckheere N, Almeida R et al (2003) Human MUC2 mucin gene is transcriptionally regulated by cdx homeodomain proteins in gastrointestinal carcinoma cell lines. J Biol Chem 278:51549–51556. https://doi.org/10.1074/jbc.M309019200

Michailidou K, Beesley J, Lindstrom S et al (2015) Genome-wide association analysis of more than 120,000 individuals identifies 15 new susceptibility loci for breast cancer. Nat Genet 47:373–380. https://doi.org/10.1038/ng.3242

Milne AN, Offerhaus GJA (2010) Early-onset gastric cancer: learning lessons from the young. World J Gastrointest Oncol 2:59–64. https://doi.org/10.4251/wjgo.v2.i2.59

Milne ANA, Carvalho R, Morsink FM et al (2006) Early-onset gastric cancers have a different molecular expression profile than conventional gastric cancers. Mod Pathol 19:564–572. https://doi.org/10.1038/modpathol.3800563

Mitchell JS, Li N, Weinhold N et al (2016) Genome-wide association study identifies multiple susceptibility loci for multiple myeloma. Nat Commun 7:12050. https://doi.org/10.1038/ncomms12050

Mizoshita T, Inada K, Tsukamoto T et al (2001) Expression of Cdx1 and Cdx2 mRNAs and relevance of this expression to differentiation in human gastrointestinal mucosa–with special emphasis on participation in intestinal metaplasia of the human stomach. Gastric Cancer 4:185–191

Mizoshita T, Tsukamoto T, Nakanishi H et al (2003) Expression of Cdx2 and the phenotype of advanced gastric cancers: relationship with prognosis. J Cancer Res Clin Oncol 129:727–734. https://doi.org/10.1007/s00432-003-0499-6

Moelans CB, Milne AN, Morsink FH et al (2011) Low frequency of HER2 amplification and overexpression in early onset gastric cancer. Cell Oncol 34:89–95. https://doi.org/10.1007/s13402-011-0021-0

Mukaisho K, Nakayama T, Hagiwara T et al (2015) Two distinct etiologies of gastric cardia adenocarcinoma: interactions among pH, Helicobacter pylori, and bile acids. Front Microbiol 6:412. https://doi.org/10.3389/fmicb.2015.00412

Nasri S, More H, Graziano F et al (2008) A novel diffuse gastric cancer susceptibility variant in E-cadherin (CDH1) intron 2: a case control study in an Italian population. BMC Cancer 8:138. https://doi.org/10.1186/1471-2407-8-138

Normanno N, Bianco C, Strizzi L et al (2005) The ErbB receptors and their ligands in cancer: an overview. Curr Drug Targets 6:243–257

Ohki R, Nemoto J, Murasawa H et al (2000) Reprimo, a new candidate mediator of the p53-mediated cell cycle arrest at the G2 phase. J Biol Chem 275:22627–22630. https://doi.org/10.1074/jbc.C000235200

Oliveira C, Ferreira P, Nabais S et al (2004) E-cadherin (CDH1) and p53 rather than SMAD4 and caspase-10 germline mutations contribute to genetic predisposition in Portuguese gastric cancer patients. Eur J Cancer 40:1897–1903. https://doi.org/10.1016/j.ejca.2004.04.027

Orlando G, Law PJ, Palin K et al (2016) Variation at 2q35 (PNKD and TMBIM1) influences colorectal cancer risk and identifies a pleiotropic effect with inflammatory bowel disease. Hum Mol Genet 25:2349–2359. https://doi.org/10.1093/hmg/ddw087

Ou S-HI (2012) Second-generation irreversible epidermal growth factor receptor (EGFR) tyrosine kinase inhibitors (TKIs): a better mousetrap? A review of the clinical evidence. Crit Rev Oncol Hematol 83:407–421. https://doi.org/10.1016/j.critrevonc.2011.11.010

Park WS, Oh RR, Park JY et al (2000) Mapping of a new target region of allelic loss at 21q22 in primary gastric cancers. Cancer Lett 159:15–21

Petersen GM, Amundadottir L, Fuchs CS et al (2010) A genome-wide association study identifies pancreatic cancer susceptibility loci on chromosomes 13q22.1, 1q32.1 and 5p15.33. Nat Genet 42:224–228. https://doi.org/10.1038/ng.522

Pirini F, Noazin S, Jahuira-Arias MH et al (2017) Early detection of gastric cancer using global, genome-wide and IRF4, ELMO1, CLIP4 and MSC DNA methylation in endoscopic biopsies. Oncotarget 8:38501–38516. https://doi.org/10.18632/oncotarget.16258

Plummer M, Franceschi S, Vignat J et al (2015) Global burden of gastric cancer attributable to Helicobacter pylori. Int J Cancer 136:487–490. https://doi.org/10.1002/ijc.28999

Rafnar T, Sulem P, Stacey SN et al (2009) Sequence variants at the TERT-CLPTM1L locus associate with many cancer types. Nat Genet 41:221–227. https://doi.org/10.1038/ng.296

Retterspitz MF, Mönig SP, Schreckenberg S et al (2010) Expression of β-catenin, MUC1 and c-Met in diffuse-type gastric carcinomas: correlations with tumour progression and prognosis. Anticancer Res 30:4635–4641

Risinger JI, Berchuck A, Kohler MF, Boyd J (1994) Mutations of the E–cadherin gene in human gynecologic cancers. Nat Genet 7:98–102. https://doi.org/10.1038/ng0594-98

Rothman N, Garcia-Closas M, Chatterjee N et al (2010) A multi-stage genome-wide association study of bladder cancer identifies multiple susceptibility loci. Nat Genet 42:978–984. https://doi.org/10.1038/ng.687

Sankpal NV, Mayo MW, Powell SM (2005) Transcriptional repression of TFF1 in gastric epithelial cells by CCAAT/enhancer binding protein-beta. Biochim Biophys Acta 1728:1–10. https://doi.org/10.1016/j.bbaexp.2004.12.014

Scelo G, Purdue MP, Brown KM et al (2017) Genome-wide association study identifies multiple risk loci for renal cell carcinoma. Nat Commun 8:15724. https://doi.org/10.1038/ncomms15724

Seno H, Oshima M, Taniguchi M-A et al (2002) CDX2 expression in the stomach with intestinal metaplasia and intestinal-type cancer: prognostic implications. Int J Oncol 21:769–774

Servetas SL, Bridge DR, Merrell DS (2016) Molecular mechanisms of gastric cancer initiation and progression by Helicobacter pylori. Curr Opin Infect Dis 29:304–310. https://doi.org/10.1097/QCO.0000000000000248

Shapiro L, Fannon AM, Kwong PD et al (1995) Structural basis of cell-cell adhesion by cadherins. Nature 374:327–337. https://doi.org/10.1038/374327a0

Siegel RL, Miller KD, Jemal A (2018) Cancer statistics, 2018. CA Cancer J Clin 68:7–30. https://doi.org/10.3322/caac.21442

Sineva GS, Pospelov VA (2014) Chapter 2—β-catenin in pluripotency: adhering to self-renewal or Wnting to differentiate? In: Jeon KW (ed) International review of cell and molecular biology. Academic, Cambridge, pp 53–78

Sitarz R, Skierucha M, Mielko J et al (2018) Gastric cancer: epidemiology, prevention, classification, and treatment. Cancer Manag Res 10:239–248. https://doi.org/10.2147/CMAR.S149619

Skibola CF, Berndt SI, Vijai J et al (2014) Genome-wide association study identifies five susceptibility loci for follicular lymphoma outside the HLA region. Am J Hum Genet 95:462–471. https://doi.org/10.1016/j.ajhg.2014.09.004

Skierucha M, Milne AN, Offerhaus GJA et al (2016) Molecular alterations in gastric cancer with special reference to the early-onset subtype. World J Gastroenterol 22:2460–2474. https://doi.org/10.3748/wjg.v22.i8.2460

Soutto M, Belkhiri A, Piazuelo MB et al (2011) Loss of TFF1 is associated with activation of NF-κB–mediated inflammation and gastric neoplasia in mice and humans. J Clin Invest 121:1753–1767. https://doi.org/10.1172/JCI43922

Sud A, Kinnersley B, Houlston RS (2017) Genome-wide association studies of cancer: current insights and future perspectives. Nat Rev Cancer 17:692–704. https://doi.org/10.1038/nrc.2017.82

Suh E, Chen L, Taylor J, Traber PG (1994) A homeodomain protein related to caudal regulates intestine-specific gene transcription. Mol Cell Biol 14:7340–7351

Suriani R, Colozza M, Cardesi E et al (2008) CagA and VacA Helicobacter pylori antibodies in gastric cancer. Can J Gastroenterol 22:255–258

Suriano G, Oliveira C, Ferreira P et al (2003) Identification of CDH1 germline missense mutations associated with functional inactivation of the E-cadherin protein in young gastric cancer probands. Hum Mol Genet 12:575–582

Thim L, May FEB (2005) Structure of mammalian trefoil factors and functional insights. Cell Mol Life Sci CMLS 62:2956–2973. https://doi.org/10.1007/s00018-005-5484-6

Tomita H, Takaishi S, Menheniott TR et al (2011) Inhibition of gastric carcinogenesis by the hormone gastrin is mediated by suppression of TFF1 epigenetic silencing. Gastroenterology 140:879–891. https://doi.org/10.1053/j.gastro.2010.11.037

Turnbull C, Perdeaux ER, Pernet D et al (2012) A genome-wide association study identifies susceptibility loci for Wilms tumor. Nat Genet 44:681–684. https://doi.org/10.1038/ng.2251

van der Woude CJ, Kleibeuker JH, Tiebosch ATGM et al (2003) Diffuse and intestinal type gastric carcinomas differ in their expression of apoptosis related proteins. J Clin Pathol 56:699–702

Virgilio E, Giarnieri E, Giovagnoli MR et al (2018) Gastric juice microRNAs as potential biomarkers for screening gastric cancer: a systematic review. Anticancer Res 38:613–616

Wang Y, Broderick P, Webb E et al (2008) Common 5p15.33 and 6p21.33 variants influence lung cancer risk. Nat Genet 40:1407–1409. https://doi.org/10.1038/ng.273

Wang L-K, Hsiao T-H, Hong T-M et al (2014) MicroRNA-133a suppresses multiple oncogenic membrane receptors and cell invasion in non-small cell lung carcinoma. PLoS One 9:e96765. https://doi.org/10.1371/journal.pone.0096765

Wang Z, McGlynn KA, Rajpert-De Meyts E et al (2017) Meta-analysis of five genome-wide association studies identifies multiple new loci associated with testicular germ cell tumor. Nat Genet 49:1141–1147. https://doi.org/10.1038/ng.3879

Wichmann IA, Zavala K, Hoffmann FG et al (2016) Evolutionary history of the reprimo tumor suppressor gene family in vertebrates with a description of a new reprimo gene lineage. Gene 591:245–254. https://doi.org/10.1016/j.gene.2016.07.036

Wolpin BM, Rizzato C, Kraft P et al (2014) Genome-wide association study identifies multiple susceptibility loci for pancreatic cancer. Nat Genet 46:994–1000. https://doi.org/10.1038/ng.3052

Wu J-H, Yao Y-L, Gu T et al (2014) MiR-421 regulates apoptosis of BGC-823 gastric cancer cells by targeting caspase-3. Asian Pac J Cancer Prev APJCP 15:5463–5468

Yamazaki K, Onouchi Y, Takazoe M et al (2007) Association analysis of genetic variants in IL23R, ATG16L1 and 5p13.1 loci with Crohn's disease in Japanese patients. J Hum Genet 52:575–583. https://doi.org/10.1007/s10038-007-0156-z

Yang P, Zhang M, Liu X, Pu H (2017) MicroRNA-421 promotes the proliferation and metastasis of gastric cancer cells by targeting claudin-11. Exp Ther Med 14:2625–2632. https://doi.org/10.3892/etm.2017.4798

Yi M, Li M, Long X et al (2016) miR-520e regulates cell proliferation, apoptosis and migration in breast cancer. Oncol Lett 12:3543–3548. https://doi.org/10.3892/ol.2016.5085
Yu X, Luo L, Wu Y et al (2013) Gastric juice miR-129 as a potential biomarker for screening gastric cancer. Med Oncol 30:365. https://doi.org/10.1007/s12032-012-0365-y
Zhang Y, Liu X, Fan Y et al (2006) Germline mutations and polymorphic variants in MMR, E-cadherin and MYH genes associated with familial gastric cancer in Jiangsu of China. Int J Cancer 119:2592–2596. https://doi.org/10.1002/ijc.22206
Zhong X, Yu J, Zhang W et al (2012) MicroRNA-421 functions as an oncogenic miRNA in biliary tract cancer through down-regulating farnesoid X receptor expression. Gene 493:44–51. https://doi.org/10.1016/j.gene.2011.11.028
Zhou Y-N, Xu C-P, Han B et al (2002) Expression of E-cadherin and β-catenin in gastric carcinoma and its correlation with the clinicopathological features and patient survival. World J Gastroenterol 8:987–993. https://doi.org/10.3748/wjg.v8.i6.987

Chapter 12
Big Data Analytics and Radiomics to Discover Diagnostics and Therapeutics for Gastric Cancer

Kummetha Jagadish, B. Pratap Naidu, G. Mohana Sheela, Nageswara Rao Reddy Neelapu, and Pallaval Veera Bramhachari

Abstract Cancer is the cause of early death and it is unique. A cancer diagnosis is complicated, and treatment outcomes vary from patient to patient. Improving cancer diagnosis may help in early diagnosis and reduces early deaths. The most common method for the diagnosis of gastrointestinal cancer is gastroscopic imaging. The availability of white light, non-magnifying images, and manual pathological examination are the major drawbacks of the system. Imaging methods like X-Ray, Computed Tomography (CT), Magnetic Resonance Imaging (MRI), Nuclear Medicine (NM) Positron Emission Tomography (PET), and Ultrasound (US) had revolutionized the diagnosis of gastrointestinal cancer. The disadvantage with these radiological images is that they contain more information and content, which is not visible to the clinician's eye. Radiomics is a process of conversion of digital medical images into mineable high-dimensional data. In this chapter, the use of big data in radiomics as a tool for gastrointestinal cancer diagnosis and prognosis is discussed. This provides information and helps in the early detection of gastrointestinal cancer.

Keywords Cancer · Gastrointestinal cancer · Radiological images · Radiomics

K. Jagadish
GVK BIO, Hyderabad, Telangana State, India

B. P. Naidu (✉) · G. M. Sheela · P. Veera Bramhachari
Department of Biotechnology, Krishna University, Machilipatnam, Andhra Pradesh, India

N. R. R. Neelapu
Department of Biochemistry and Bioinformatics, GITAM Institute of Science, Gandhi Institute of Technology and Management (GITAM) (Deemed-to-be-University), Visakhapatnam, Andhra Pradesh, India

P. Veera Bramhachari, N. R. R. Neelapu (eds.), *Recent Advancements in Biomarkers and Early Detection of Gastrointestinal Cancers*, Diagnostics and Therapeutic Advances in GI Malignancies, https://doi.org/10.1007/978-981-15-4431-6_12

12.1 Introduction

Gastric cancer (GC) is the third-highest based on lethality and fourth-highest based on morbidity of all cancers (Rawla and Barsouk 2019). According to WHO GLOBOCAN statistics ~1,033,701 new cases and ~782,000 deaths were recorded for gastric cancer in 2018 (Bray et al. 2018). The most important reason for the situation is the lack of methods for early diagnosis. Early signs of GC are extremely difficult to detect, often bearing a close resemblance to inflammation. When diagnosed, half of the GC patients are in an advanced stage having a 5-year survival rate which is lower than 30% (Rugge et al. 2014). Early detection and proper treatment following precise risk classification are crucial for improving the outcome of gastric cancer. Gastroscopic imaging is a widely used method for the diagnosis of GC. The drawbacks include the availability of white light and non-magnifying images, manual pathological inspection, the inability of the human eye to identify minor lesions from the images, requirement of high-quality, narrow-band imaging (or laser-based), and requirement of magnified images for present image reading algorithms. Recently, computer-aided methods are expected to play an important role in the detection of GC. Development of advanced magnifying endoscopes, deep learning methods, and machine learning methods; and availability of histopathological images enabled reading the weakly labeled images. These advances and developments improved the diagnosis of GC (Ronald 2018). The above methods are used to diagnose GC, but the methods to diagnose GC at very early stages are required.

GC is a disease that evolves due to various genetic and epigenetic alterations. GC originates due to the sequential accumulation of molecular and genetic alterations in stomach epithelial cells. Multidisciplinary diagnostic approaches integrating endoscopy, serology, histology, and molecular profiling are the appropriate approaches for stratification of patients into different GC risk classes. Big data analytics and machine learning methods can bring together the above-mentioned multiple disciplinary diagnostics to help in early diagnosis of GC. The term "big data" refers "to huge amounts of information that can be analyzed by high-performance computers to reveal patterns, trends, and associations." In medical terms, big data includes clinical and genomic data that is derived from patients during diagnostic testing and treatment. Big data analytics can reveal the patterns and relationships among a large amount of data in a single or several data sets. The data analytics uses several techniques like statistics and artificial intelligence to reveal the hidden patterns and rules in big data. Big data analytics is used in a variety of activities or applications, and the application of big data analytics to the gastric cancer diagnosis is an upcoming trend. Recent advances in understanding the molecular mechanisms that mediate GC and big data analytics were promising and paved the path for the development of more effective diagnosis strategies. Extensive research is also carried out in the field of image analysis for diagnosing and identifying GC at the early stages. Recently, Japanese research group successfully used artificial intelligence to diagnose GC (In breakthrough, Japanese researchers use AI to identify early-stage stomach cancer with high accuracy 2018). In this chapter, the recent updates on the role of big data in radiomics, machine learning, or artificial intelligence for early diagnosis of GC are discussed.

12.2 Radiomics

Radiological imaging techniques are powerful noninvasive tools used for detection, differentiation, and diagnosis of different tissue characteristics in patients. Radiologists acquire a huge amount of data by imaging tissues from various views and angles for complete image phenotypes. The imaging methods include X-Ray, Computed Tomography (CT), Magnetic Resonance Imaging (MRI), Nuclear Medicine (NM), Positron Emission Tomography (PET), and Ultrasound (US). Each of these modalities creates tissue contrast based on the differences in the tissue between normal or abnormal. These tissue contrasts are exploited by the radiologist to identify patterns for diagnosis. Radiologists are trained to understand the imaging phenotypes and transcribe these observations to correlate with underlying diseases. Traditionally, these medical images are treated as pictures intended solely for visual interpretation. However, each of the radiological images contains more information content not visible to the clinician's eye. This "hidden" information creates a "radiological texture" which can provide much more information about the tissue of interest than previously thought.

Radiomics is a promising field of medical research that employs a combination of computer-aided deep learning methods and human skills to convert digital medical images into mineable high-dimensional data (Lambin et al. 2012). Then translates the metrics obtained from texture and other features on radiological images. Figure 12.1 represents the basic workflow of radionics with imaging, segmentation, feature extraction, and analysis. Medical or radiological images are generated from various modalities such as X-Ray, CT, MRI, PET, and US. Segmentation is performed to define the tumor region on the radiological images. Then, radiomics employs machine learning methods to extract huge quantities of imaging features like tumor intensity, texture, and shape from radiological images. Radiomics features contain useful spatial and textural information on the grayscale patterns and the correlation between image pixels. These features can be modeled or used for analysis assessed for their prognostic power, or linked with stage, or gene expression (Parekh

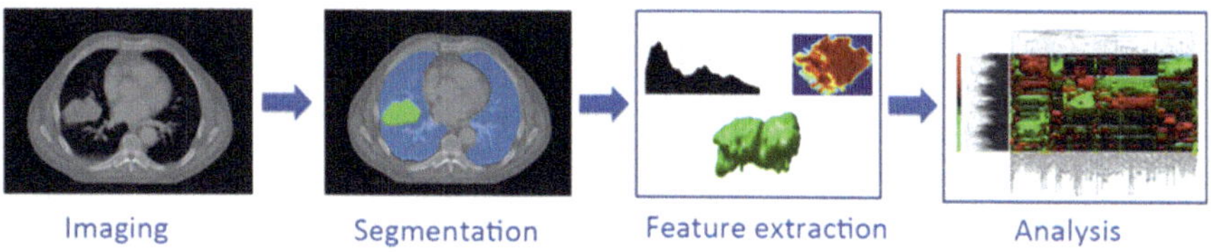

Fig. 12.1 The workflow of radiomics, imaging, segmentation, feature extraction, and analysis of features for prognosis or diagnosis (Parekh and Jacobs 2016)

and Jacobs 2016). Then, this can supplement as an adjunct instrument to discover or predict or decode concealed genetic and molecular traits for decision support, individualized diagnosis, and treatment guidance (Kumar et al. 2012a; Cook et al. 2014; Court et al. 2016; Gillies et al. 2016; Narang et al. 2016; Yip and Aerts 2016; Sala et al. 2017). This information obtained from these radiological images can also be combined with additional OMICS data (genomics, proteomics, metabolomics, and transcriptomics) for further analysis. This branch of study is called radiogenomics, and this is an upcoming technology that is applied for diagnosis and prognosis of multiple cancers (Zinn et al. 2011).

12.3 Big Data in Radiomics for Diagnosis and Prognosis of Gastric Cancer

Radiomics is an upcoming technology that can be used in general to predict gastric cancer and plan for a course of treatment. Recently, multiple studies have discussed the possibility of using radiomics and artificial intelligence for diagnosis and prognosis of gastric cancer (Jiang et al. 2018a, b; Li et al. 2018a, b; Keek et al. 2018; Acharya et al. 2018).

12.3.1 Radiomics in Preoperative Prediction of Lymph Node Metastasis

Feng et al. (2019) developed and validated an automatic decision support system (DSS) for preoperative reporting of the risk for lymph node metastasis in GC. The clinical and imaging data were analyzed using a machine learning-based approach. The clinical, pathological, and CT imaging data of 490 patients diagnosed with GC was collected. Standard gastric contrast-enhanced CT scans of the same patients were also obtained within 10 days of surgery and all gastric CT studies were performed using a 64-slice scanner. Of the 490 patients, 297 were reported with LN metastasis and also with the metastatic rate of 60.6%. Thirteen relevant radiomics features were selected, ranked, and modeled using a support vector machine (SVM) classifier based on 326 training and validation data sets. A model test was performed independently with a test set size (n) 164. The comparison was made between the Clinical Decision Support System (CDSS) and the conventional staging criterion performed by two expert radiologists for the diagnostic performance of CDSS. The DSS was better able to predict LN metastasis (accuracy 76.4%) than the conventional staging (accuracy 71.3%). Automatic DSS employing SVM classifier was able to predict LN status in patients with GC based on 13 radiomics features.

12.3.2 Radiomics to Predict Prognosis and Benefit from Chemotherapy

Jiang et al. (2018b) developed and validated a radiomics signature for the prediction of gastric cancer and the benefits of chemotherapy. A sample of 1591 patients histologically confirmed with gastric adenocarcinoma, and standard unenhanced and contrast-enhanced abdominal CT performed within 30 days was included for analysis. Radiomics signature and radiomics nomograms were generated from the sample. The lasso-cox regression model was performed on 228 patients to generate radiomics signature based on 19 selected features. Radiomics nomograms integrated with radiomics signature were constructed on TNM staging. Radiomics signature was able to predict patients with stage II and III of GC and may benefit chemotherapy.

Both the case studies could predict gastric cancer with very high accuracy. Though good progress is seen with radiomics, some challenges also exist. The major challenges to be addressed for analyzing the data are standard image acquisition methods, image reconstruction methods, optimized algorithmic approaches, and statistical approaches. Databases like the National Biomedical Imaging Archive (NBIA) (Nicholas et al. 2012), Cancer Imaging Program (CIP) (Dobranowski et al. 2014), and The Cancer Imaging Archive (TCIA) (Clark et al. 2013) are available. Developing integrated radiomics images with defined rules might help in addressing the challenges with image acquisition (Kumar et al. 2012b). However, these computer- assisted clinical decision-making methods require further external, multicenter, and evidence-based validation. Further, these applications may serve as a tool for personalized diagnosis and guidance for treatment (Parekh and Jacobs 2016).

12.4 Conclusion and Future Perspectives

In conclusion, radiomics analysis uses a machine-learning approach to provide an alternative to conventional radiologic methods. This, in turn, changes the facet of clinical decision-making of the present and future generations of patients suffering from gastric cancer. The future direction of radiomics includes correlating and integrating OMICS data with radiomics features extracted from radiological images and integrate them to create a more efficient and robust prognostic model. This, well aid clinicians in regular practice, personalized medicine, and paves new direction for the cancer diagnosis and prognosis (Mazurowski 2015; Rutman and Kuo 2009).

Acknowledgments KVJ is grateful to GVK BIO, Hyderabad, for providing necessary facilities to carry out the research work and for extending constant support. NNRR is grateful to GITAM (Deemed-to-be-University) for providing necessary facilities to carry out the research work and for extending constant support. Prathap Naidu, Dr. Mohana Sheela, and Dr. PVBC are grateful to Krishna University for providing the necessary facilities to carry out the research work and for extending constant support.

Conflict of Interest Statement The authors declare that there is no potential conflict of interest.

References

Acharya UR, Hagiwara Y, Sudarshan VK, Chan WY, Ng KH (2018) Towards precision medicine: from quantitative imaging to radiomics. J Zhejiang Univ Sci B 19(1):6–24. https://doi.org/10.1631/jzus.B1700260

Bray F, Ferlay J, Soerjomataram I, Siegel RL, Torre LA, Jemal A (2018) Global cancer statistics 2018: GLOBOCAN estimates of incidence and mortality worldwide for 36 cancers in 185 countries. CA Cancer J Clin 68(6):394–424

Clark K, Vendt B, Smith K, Freymann J, Kirby J, Koppel P, Moore S, Phillips S, Maffitt D, Pringle M, Tarbox L, Prior F (2013) The Cancer Imaging Archive (TCIA): maintaining and operating a public information repository. J Digit Imaging 26(6):1045–1057. https://www.cancerimagingarchive.net/

Cook GJR, Siddique M, Taylor BP, Yip C, Chicklore S, Goh V (2014) Radiomics in PET: principles and applications. Clin Transl Imaging 2(3):269–276

Court LE, Fave X, Mackin D, Lee J, Yang J, Zhang L (2016) Computational resources for radiomics. Transl Cancer Res 5(4):340–348

Dobranowski J, Melamed S, Langer D, Bedford C (2014) The cancer imaging program quality framework at cancer care Ontario: the first five years. J Clin Oncol 32(30_suppl):244. https://imaging.cancer.gov/

Feng QX, Liu C, Qi L, Sun SW, Song Y, Yang G, Zhang YD, Liu XS (2019) An intelligent clinical decision support system for preoperative prediction of lymph node metastasis in gastric cancer. J Am Coll Radiol 16(7):952–960

Gillies RJ, Kinahan PE, Hricak H (2016) Radiomics: images are more than pictures, they are data. Radiology 278(2):563–577

In breakthrough, Japanese researchers use AI to identify early-stage stomach cancer with high accuracy (2018). https://www.japantimes.co.jp/news/2018/07/22/national/science-health/japanese-researchers-use-ai-identify-early-stage-stomach-cancer-high-accuracy/#.XFqIPFwza70. Accessed 12 Oct 2019

Jiang Y, Chen C, Xie J, Wang W, Zha X, Lv W, Chen H, Hu Y, Li T, Yu J, Zhou Z, Xu Y, Li G (2018a) Radiomics signature of computed tomography imaging for prediction of survival and chemotherapeutic benefits in gastric cancer. EBioMedicine 36:171–182

Jiang Y, Yuan Q, Lv W, Xi S, Huang W, Sun Z, Chen H, Zhao L, Liu W, Hu Y, Lu L, Ma J, Li T, Yu J, Wang Q, Li G (2018b) Radiomic signature of 18F fluorodeoxyglucose PET/CT for prediction of gastric cancer survival and chemotherapeutic benefits. Theranostics 8 (21):5915–5928

Keek SA, Leijenaar RT, Jochems A, Woodruff HC (2018) A review on radiomics and the future of theranostics for patient selection in precision medicine. Br J Radiol 91(1091):20170926. https://doi.org/10.1259/bjr.20170926

Kumar V, Gu Y, Basu S, Kumar V, Gu Y, Basu S, Berglund A, Eschrich SA, Schabath MB, Forster K, Aerts HJ, Dekker A, Fenstermacher D, Goldgof DB, Hall LO, Lambin P, Balagurunathan Y, Gatenby RA, Gillies RJ (2012a) Radiomics: the process and the challenges. Magn Reson Imaging 30(9):1234–1248

Kumar V, Gu Y, Basu S, Berglund A, Eschrich SA, Schabath MB, Forster K, Aerts HJ, Dekker A, Fenstermacher D, Goldgof DB, Hall LO, Lambin P, Balagurunathan Y, Gatenby RA, Gillies RJ (2012b) Radiomics: the process and the challenges. Magn Reson Imaging 30(9):1234–1248. https://doi.org/10.1016/j.mri.2012.06.010

Lambin P, Rios-Velazquez E, Leijenaar R, Carvalho S, van Stiphout RG, Granton P, Zegers CM, Gillies R, Boellard R, Dekker A, Aerts HJ (2012) Radiomics: extracting more information from medical images using advanced feature analysis. Eur J Cancer 48(4):441–446

Li Z, Zhang D, Dai Y, Dong J, Wu L, Li Y, Cheng Z, Ding Y, Liu Z (2018a) Computed tomography-based radiomics for prediction of neoadjuvant chemotherapy outcomes in locally advanced gastric cancer: a pilot study. Chin J Cancer Res 30(4):406–414

Li W, Zhang L, Tian C, Song H, Fang M, Hu C, Zang Y, Cao Y, Dai S, Wang F, Dong D, Wang R, Tian J (2018b) Prognostic value of computed tomography radiomics features in patients with gastric cancer following curative resection. J Eur Radiol 29(6):3079–3089

Mazurowski MA (2015) Radiogenomics: what it is and why it is important. J Am Coll Radiol 12 (8):862–866

Narang S, Lehrer M, Yang D, Lee J, Rao A (2016) Radiomics in glioblastoma: current status, challenges and potential opportunities. Transl Cancer Res 5(4):383–397

Nicholas A, Mulhern P, Siegel E (2012) The National Biomedical Imaging Archive: a repository of advanced imaging information. J Nucl Med 53(Suppl 1):1009. https://imaging.nci.nih.gov/ncia/login.jsf

Parekh V, Jacobs MA (2016) Radiomics: a new application from established techniques. Expert Rev Precis Med Drug Dev 1(2):207–226

Rawla P, Barsouk A (2019) Epidemiology of gastric cancer: global trends, risk factors and prevention. Prz Gastroenterol 14(1):26–38

Ronald S (2018) The impact of deep learning and artificial intelligence on radiology. In: 1st conference on medical imaging with deep learning (MIDL 2018), Amsterdam, The Netherlands, pp 6

Rugge M, Capelle LG, Fassan M (2014) Individual risk stratification of gastric cancer: evolving concepts and their impact on clinical practice. Best Pract Res Clin Gastroenterol 28 (6):1043–1053

Rutman AM, Kuo MD (2009) Radiogenomics: creating a link between molecular diagnostics and diagnostic imaging. Eur J Radiol 70(2):232–241

Sala E, Mema E, Himoto Y, Veeraraghavan H, Brenton JD, Snyder A, Weigelt B, Vargas HA (2017) Unravelling tumour heterogeneity using next-generation imaging: radiomics, radiogenomics, and habitat imaging. Clin Radiol 72(1):3–10

Yip SSF, Aerts HJWL (2016) Applications and limitations of radiomics. Phys Med Biol 61(13): R150–R166

Zinn PO, Mahajan B, Sathyan P, Singh SK, Majumder S, Jolesz FA, Colen RR (2011) Radiogenomic mapping of edema/cellular invasion MRI-phenotypes in glioblastoma multiforme. PLoS One 6(10):e25451

Chapter 13
Systems Biology Approach for Early Prognosis of Gastrointestinal Cancer

Pavani Sanapala and Sudhakar Pola

Abstract In order to treat cancer as a disease of the phenome, initially, we have to understand the functional difference that takes place from a normal cell. Cancer systems biology research is necessary to analyze the complexities of various pathways involving signaling, regulation of the gene, cell metabolism, and alterations in its system caused due to mutations leading to malignancy. As these approaches seem complicated with several interlinks connecting pathways, it is necessary to signify it in the form of a computational model. Gastrointestinal cancer has high death rates all around the world; hence, there is a necessitate to identify suitable biomarkers for GC. This chapter is to illustrate tools for early diagnosis of gastrointestinal cancer through high-throughput systems biology approaches.

Keywords Gastrointestinal cancer · Systems biology · Genetic factors · High-throughput techniques · Metabolomics

13.1 Introduction

Approximately out of 57 million people deaths worldwide, cancer is the most avertible and curable chronic disease with a high incidence of mortality and morbidity (Albreht et al. 2008). Gastrointestinal cancer or the GI cancer is together termed to a set of cancers affecting the digestive system. Cancers, namely the gallbladder and biliary tract cancer, esophagus cancer, liver, pancreas, stomach, small and large intestine cancers, colon, rectum, and anus together constitute the GI cancers. According to GLOBOCON 2012, of all the GI cancers, the stomach, liver, and colorectal are major cancers affected, as a result, more focus and attention are drawn for their management (Aman and Buzdar 2005; Azizi et al. 2000). Statistics prevalence of GI cancer all around the world is placed fifth as the most

P. Sanapala · S. Pola (✉)
Department of Biotechnology, Andhra University, Visakhapatnam, Andhra Pradesh, India
e-mail: sudhakar@andhrauniversity.edu.in

P. Veera Bramhachari, N. R. R. Neelapu (eds.), *Recent Advancements in Biomarkers and Early Detection of Gastrointestinal Cancers*, Diagnostics and Therapeutic Advances in GI Malignancies, https://doi.org/10.1007/978-981-15-4431-6_13

common cancers and the third cause for fatalities (Anvar et al. 2018). The effect of GI cancer shows no discrimination among feminine and masculine. Furthermore, due to high metastatic reappearance in the advanced gastric cancer stage, the prognosis of patients remained underprivileged (Macdonald 2006; Cunningham and Chua 2007), as well the metastasis mechanism underlying the major cause of cancer is not well differentiated (Rajdev 2010; Yilmaz and Christofori 2010).

Besides non-genetic factors, genetic factors (proteins) play a crucial role in cancer pathogenesis and also raise the expression levels in gastric cancer. Studies recently have given common pathways in various GI cancers, which assists in introducing capable treatments to manage cancer (Guo and Jiang 2009; Wu and Qu 2015). Furthermore, examining the expression patterns of these biomarkers aids in the early diagnosis of GI cancer patients with no initial symptoms. However, to date, the use of endoscopy for early diagnosis of human gastric cancer is widely applied (Tashiro et al. 2006; Sipponen et al. 2002; Lu et al. 2008) even though of its contradictory investigative competency. Currently, cancer research has advantaged from accessible high-throughput gene expression datasets (Serrano 2007; Fehrmann et al. 2015). This, in turn, has increased the opportunities of discovering novel cancer biomarkers that are used for calculating the risk of disease, screening and monitoring of cancer at different stages, and also in detecting the disease reappearance (Malley and Pidgeon 2016).

13.2 Systems Biology

Over the past decade, systems biology in life science research has surfaced as an influential novel model. The entire structure has emerging functions that are not able to be seen at the level of the parts. Systems biology classifies and evaluates the interactions internally of all the elements in the operating system to comprehend the system working (Kang et al. 2016). In biological structure, systems biology goals for analyzing thousands of genes or proteins locally or globally, investigate several convolutions of molecules, complexes, networks, and many more scrutinize network modeling likely protein–protein interaction, cell signaling, metabolism, and gene regulatory systems and also detect functional effects by sequential, environment, genetic and epigenetic changes (Koutsogiannouli et al. 2013).

Progress in this field has enhanced the capability to gather valuable information from datasets of highly advanced technology and biomedical investigations. To investigate the disease pertinent gene prioritization, several computational looms such as functional annotation, sequence-based knowledge, comparison of phenotype have been signified as the applications of systems biology approaches. The biology paradigm of the systems uses mathematical models as a significant feature in studying the framework of network structures and dynamics (Pradhan et al. 2012).

Pooling data for a vast number of species together with the information computing gene expression and the intensity of macromolecules and metabolites unlock the prospect of acquiring network-level data. However, this type of information or data

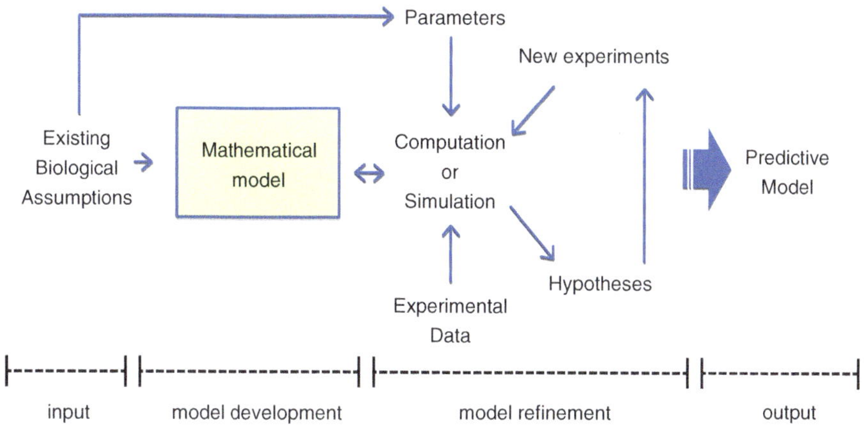

Fig. 13.1 The systems biology stratagem (Laubenbacher et al. 2009)

makes it possible in discovering novel approaches, which is an ultimate aim of a systems biology approach towards cancer where an appropriate biological network aiding with a mathematical approach is to be valued (Loos and Schadt 2012). This approach of strategies with several steps is illustrated in Fig. 13.1. This illustration portrays the general strategy to be followed. The initial point is the existing biological assumptions about the system that is preferred to construct the mathematical model. The approach is distinguished through biological parameters and experimental information to which the model be able to be en suite through recreation or computation. During this stage, the model generates hypotheses that can be experimentally analyzed. By a repetitive process of hypotheses, the approach is refined until confining the predictive model.

13.3 System Approaches for Cancer Biology

Cancer systems biology (CSB) identifies various disciplines and data categories that are beneficially brought to bear unaided or in the group for the analytical study of cancer. Highlights from the Conference on Systems Approaches to Cancer Biology signified the biological insights profited through synergistic relations and also conversed the unique challenges. The theme of the meeting was too aware of the young investigators in the field of systems biology approaches to cancer. Foremost the cancer complexity and heterogeneity of cancer within and across the patients is to be analyzed through computational tools, the glean approaches for preclinical and clinical datasets.

Many scientists have come forth with various statistical tools in learning cancer biology using system approaches. One such study is the use of the Hot-Net2 algorithm, which assimilates cancer mutational data with known protein interactions

(Leiserson et al. 2015). However, the study showed only a few cancer-driven mutations. However, analyzing the exome and phospho-proteomes on cell lines using the ReKINect tool showed effective towards kinases mutation. Considering the two approaches together recommended that integrating protein-level data with mutational in the sequence is a comprehensive approach but care is required when presumptuous the mutation consequence with protein activity.

In a few cases, the degree and complication of system studies make reproducibility challenges, especially with clinical and preclinical investigations (Begley and Ioannidis 2015). Implementing these challenges in research involves the facility to review data analysis and laboratory protocols, for example, network algorithms accessible on GitHub and NDEx resources (Omberg et al. 2013; Pratt et al. 2015). To recapitulate and build a prior work regarding reproducible research data sharing is a fundamental tool. Also, creating a new model or approach from existing data helps in finding novel findings.

Medical transformation and collaborative science study showed p53 as a key tumor suppressor that directs the treatment of engaging DNA damaging therapies in cancer such as cisplatin (Paek et al. 2016) and radiation (Chen et al. 2016). A deep sequencing profiling method CAPP-seq is discovered for besieged sequencing of circulating tumor DNA (ctDNA) in tracking the disease progression. These system approaches reveal many mutational and amplification changes taking place in cancer biology as an early prognosis.

Schroeder et al. gave a hopeful example of utilizing computational modeling for effectual prediction of new and non-intuitive cancer therapeutic drug targets, especially ErbB3 via in vitro and in vivo validation (Schoeberl et al. 2009).

13.4 Systems Biology and Gastrointestinal Cancers

Metastasis is a change happening in cancer from a controllable form to uncontrollable or fatal. In support of metastasis to take place, the tumor cells initially must be detached from the actual tumor, followed by relocation via blood circulation and initiate a new colony at an unusual place rather than the actual part of the organism. Genetic factors are also viewed as a determining feature in transformation. Systems biology study aspires to identify the properties and functions of a known system, which, in relative to cancer, possibly be tumor cell lines, xenografts, genetically engineered primary cells of pre- and post-treated patients.

Factors essential at the molecular level for regulation are the cell signaling molecules, catabolic enzymes, cell growth factors, and a range of angiogenesis factors. A swift in progress of high-throughput techniques facilitate the expression of a large number of genes concurrently with the involvement of cDNA and oligonucleotide microarrays. This extends an unprecedented opportunity that differentiates the fundamental mechanism of carcinogenesis. Genome-wide studies with DNA arrays have turned out to be a bastion of genomics investigations. However, these experiments habitually produce genes and proteins more than ten folds, with noisy outcomes now and then.

13.4.1 A Systems Biology Approach of Gastric Cancer Biomarkers (Genomics)

In most cases of gastric cancer, proteins play a role in pathogenesis and most frequently the expression levels of receptor tyrosine-protein kinase erbB-2 (ERBB-2) which is elevated in gastric cancer (Saito et al. 2015; Rüschoff et al. 2012). Similarly, tumor suppressor antigen p53 implicated mainly in cell division, regulation, apoptosis initiation, mutates in the majority of cancers (Starzynska et al. 1992; Fenoglio-Preiser et al. 2003; Azarhoush et al. 2008). Likewise, Gastrokine in reducing the expression of a gastrin signaling pathway can prevent gastric cancer (Uhlén et al. 2015; Kim et al. 2016; Mao et al. 2012). Besides these mentioned proteins, miR-145 performs the role of tumor suppression via the vitamin-D3-dependent pathway and also reduces its expression levels in gastric cancer (Choi et al. 2013). Examining and assembling the networks progress our understanding of the pathological mechanism in disease and also helps in identifying the disease target drugs as well as crucial diagnostic markers (Chang et al. 2015; Kann 2007).

In a study by Anvar et al. using the high-throughput method on the network and sub-networks (Fig. 13.2), identified proteins TBP-associated factor 1 (functions as reducing apoptosis and enhancing cell viability), hepatocyte nuclear factor 4 alpha (HNF4A), and TP53 had highest indices. HNF4A shows activities such as immune and stress response, programmed cell death, metabolism regulation also connecting 240 proteins in cancer-related pathways (Anvar et al. 2018). Gene set enrichment study confirmed the pathways neurotrophin signaling pathway, repair mechanism such as nucleotide excision, cell cycle, and focal adhesion add to the progression of gastric cancer. This can be explained as the rate of cell growth and cell division enlarges tumor formation stimulating cell cycle pathways in this cancer. So the author concluded this novel investigation of proteins and pathways playing a major role in the incidence of gastric cancer and be able to be the pioneer as remedial targets and significant biomarkers in gastric cancer (Safari-Alighiarloo et al. 2014).

13.4.2 Recognition of MicroRNAs as Biomarkers for Gastric Cancer by Systems Biology Approach

Gastric cancers are systems biology disease, methods for identification of markers is made difficult due to its heterogeneity and complex nature. Candidate miRNAs as a biomarker for early diagnosis using systems biology approaches were in use that distinguishes the disease patients from normal healthy controls (Yan et al. 2014).

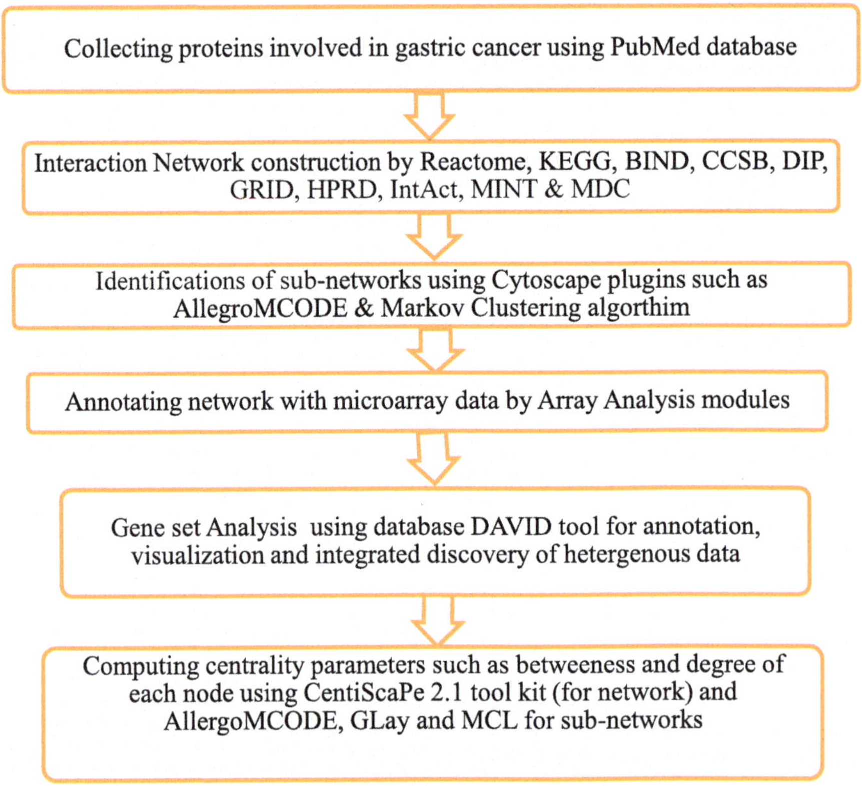

Fig. 13.2 Systems biology approach in identifying gastric cancer biomarker and pathways

This process is initiated by a collection of datasets and detection of outliner differential expressed genes with the smallest amount of ordered t-statistics, followed by refinement of candidate microRNAs using an integrative POMA (Pipeline of outliner MicroRNA analysis) method by LOSS. Later on, with the aid of heat map (by R language using "gplots") and receiver operating characteristics (ROC) curves the performance of miRNAs is evaluated. And finally, the aimed gene of candidate miRNAs was mapped to various databases likely, gene ontology, biological pathway analysis by KEGG, and meta-core pathway maps and diseases for functional enrichment analysis (Fig. 13.3).

13.4.3 Metabolomics: A Systems Biology Approach of Human Gastric Cancer

Metabolites variation between perturbed and non-perturbed computational networks gives an insight basic for disease progression, pathology, and diagnosis. So, to

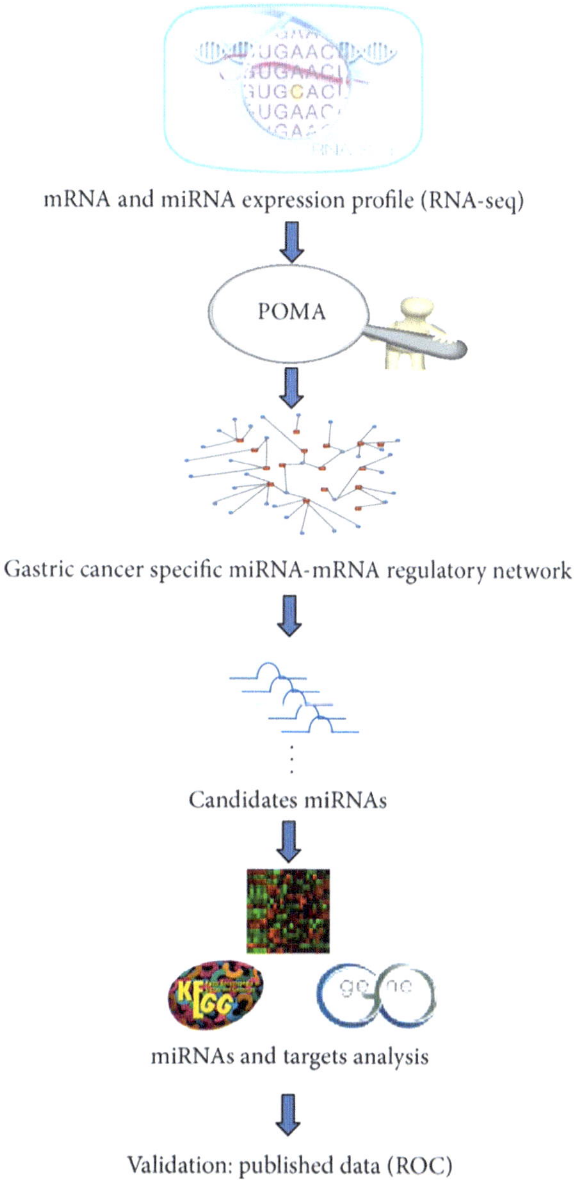

Fig. 13.3 Analysis of candidate miRNAs by systems biology approach (Yan et al. 2014)

identify and quantify these metabolites in a biological system, metabolomics study has emerged which has the ability to diagnosis the biomarkers at early-stage gastric cancer. Metabolites are not the end product but are the system genomes in interaction with the surroundings and are an integral part of the cell regulatory system (Hu et al. 2011). Metabolomics is considered as one of the new high-throughput technology (Table 13.1). Various analytical techniques are used for the study of different

Table 13.1 Overview of traditional methods and metabolomics markers of gastric cancer

State of cancer	Traditional methods	Metabolomics (Biomarkers)	References
Diagnosis	Endoscopy, biopsy	Lactic acid, butanedioic acid, citric acid, pyruvic acid, Ser, Pro	Chen et al. (2011); Ikeda et al. (2012); Holdstock and Bruce (1981); Kim et al. (2010)
Prognosis	Radiotherapy, chemotherapy surgery	Val, Ile, Ser, citrate, 3-indoxyl sulfate, Hippurate	Wu et al. (2010); Chen et al. (2010); Akagi et al. (2011)
Metastasis	Computed tomography (CT) scanning, positron emission tomography (PET)	Ala, Pro, Ser, Myo-inositol, glycerol, sarcosine	Chen et al. (2011); Layke and Lopez (2004); Song et al. (2012)
Chemosensitivity drugs	MTT Chemosensitivity assay	1-acyl-lysophosphatidylcholines and polyunsaturated fatty acids	Wang et al. (2010); Nakamura et al. (2006)

Ala Alanine, *Pro* Proline, *Ser* Serine, *Val* Valine, *Ile* Isoleucine

metabolites especially, Nuclear Magnetic Resonance (NMR), Gas Chromatography-Mass Spectroscopy (GC-MS), Liquid Chromatography-Mass Spectroscopy (LC-MS), capillary electrophoresis-mass (CE-MS) spectrophotometry, qRT-PCR (Quantitative reverse transcriptase-polymerase chain reaction), and FT-IR (Fourier Transform Infrared). The data obtained from these analyses are statistically progressed to categorize the metabolites that are differentially expressed among the samples, preferably leading to biomarker selection (Fig. 13.4).

13.5 Conclusions and Future Perspectives

This approach of systems biology made potential with high-throughput and "omics" technologies help in understanding the disease. And also identifying the proteins and pathways of gastric cancer can be useful sequentially in identifying major proteins and pathways relating to other diseases. Furthermore, coalescing metabolomics with "omics" data, a further complete understanding of the developments in cancer biology is expected to be engendered. However, besides all these findings still, further wet-lab validation is required.

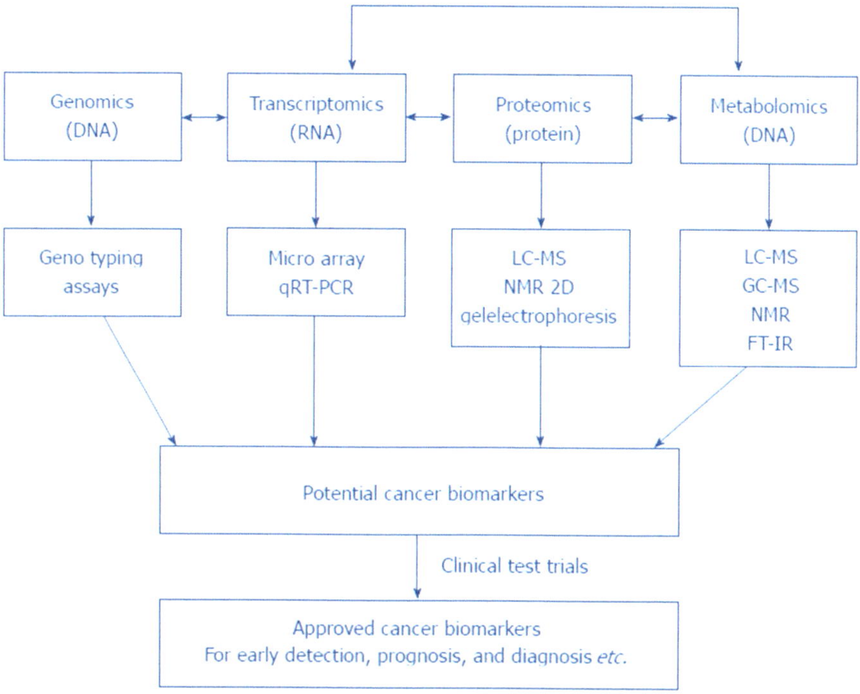

Fig. 13.4 Organization of various omics technologies (Griffin and Shockcor 2004)

Acknowledgments Pavani Sanapala and Sudhakar Pola are grateful to College of Science and Technology, Andhra University, Visakhapatnam, India, for providing necessary facilities to carry out the research work and for extending constant support.

Conflict of Interest The authors declare that there is no potential conflict of interest.

References

Akagi T, Shiraishi N, Kitano S (2011) Lymph node metastasis of gastric cancer. Cancers 3 (2):2141–2159

Albreht T, McKee M, Alexe DM, Coleman MP, Martin-Moreno JM (2008) Making progress against cancer in Europe in 2008. Eur J Cancer 44(10):1451–1456

Aman U, Buzdar R (2005) Gastrointestinal cancer. Springer, Berlin

Anvar MS, Minuchehr Z, Shahlaei M, Kheitan S (2018) Gastric cancer biomarkers; A systems biology approach. Biochem Biophys Rep 13:141–146

Azarhoush R, Keshtkar AA, Amiriani T, Kazemi-Nejad V (2008) Relationship between p53 expression and gastric cancers in cardia and antrum. Arch Iran Med 11(5):502–506

Azizi F, Hatami H, Janghorbani M (2000) Epidemiology and control of common diseases in Iran. Eshtiagh Publications, Tehran, pp 602–616

Begley CG, Ioannidis JP (2015) Reproducibility in science: improving the standard for basic and preclinical research. Circ Res 116(1):116–126

Chang SE, Gao L, Yang Y, Tong D, Guo B, Liu L et al (2015) miR-145 mediates the antiproliferative and gene regulatory effects of vitamin D3 by directly targeting E2F3 in gastric cancer cells. Oncotarget 6(10):7675

Chen JL, Tang HQ, Hu JD, Fan J, Hong J, Gu JZ (2010) Metabolomics of gastric cancer metastasis detected by gas chromatography and mass spectrometry. World J Gastroenterol 16(46):5874

Chen JL, Fan J, Tang HQ, Hu JD, Gu JZ (2011) Urinary metabolomic analysis of human gastric cancer mouse models and patients using gas chromatography/mass spectrometry. J Mol Biomark Diagn S2:003

Chen SH, Forrester W, Lahav G (2016) Schedule-dependent interaction between anticancer treatments. Science 351(6278):1204–1208

Choi BJ, Yoon JH, Choi WS, Kim O, Nam SW, Lee JY, Park WS (2013) GKN1 and miR-185 are associated with CpG island methylator phenotype in gastric cancers. Mol Cell Toxicol 9 (3):227–233

Cunningham D, Chua YJ (2007) East meets west in the treatment of gastric cancer. N Engl J Med 357(18):1863–1865

Fehrmann RS, Karjalainen JM, Krajewska M, Westra HJ, Maloney D, Simeonov A, Pers TH, Hirschhorn JN, Jansen RC, Schultes EA, van Haagen HH (2015) Gene expression analysis identifies global gene dosage sensitivity in cancer. Nat Genet 47(2):115

Fenoglio-Preiser CM, Wang J, Stemmermann GN, Noffsinger A (2003) TP53 and gastric carcinoma: a review. Hum Mutat 21(3):258–270

Griffin JL, Shockcor JP (2004) Metabolic profiles of cancer cells. Nat Rev Cancer 4(7):551

Guo W, Jiang YG (2009) Current gene expression studies in esophageal carcinoma. Curr Genomics 10(8):534–539

Holdstock G, Bruce S (1981) Endoscopy and gastric cancer. Gut 22(8):673–676

Hu JD, Tang HQ, Zhang Q, Fan J, Hong J, Gu JZ, Chen JL (2011) Prediction of gastric cancer metastasis through urinary metabolomic investigation using GC/MS. World J Gastroenterol 17 (6):727

Ikeda A, Nishiumi S, Shinohara M, Yoshie T, Hatano N, Okuno T, Bamba T, Fukusaki E, Takenawa T, Azuma T, Yoshida M (2012) Serum metabolomics as a novel diagnostic approach for gastrointestinal cancer. Biomed Chromatogr 26(5):548–558

Kang W, Cheng AS, Yu J, To KF (2016) Emerging role of Hippo pathway in gastric and other gastrointestinal cancers. World J Gastroenterol 22(3):1279

Kann MG (2007) Protein interactions and disease: computational approaches to uncover the etiology of diseases. Brief Bioinform 8(5):333–346

Kim KB, Yang JY, Kwack SJ, Park KL, Kim HS, Ryu DH, Kim YJ, Hwang GS, Lee BM (2010) Toxicometabolomics of urinary biomarkers for human gastric cancer in a mouse model. J Toxic Environ Health A 73(21–22):1420–1430

Kim O, Yoon JH, Choi WS, Ashktorab H, Smoot DT, Nam SW, Lee JY, Park WS (2016) Gastrokine 1 inhibits gastrin-induced cell proliferation. Gastric Cancer 19(2):381–391

Koutsogiannouli E, Papavassiliou AG, Papanikolaou NA (2013) Complexity in cancer biology: is systems biology the answer? Cancer Med 2(2):164–177

Laubenbacher R, Hower V, Jarrah A, Torti SV, Shulaev V, Mendes P, Torti FM, Akman S (2009) A systems biology view of cancer. Biochim Biophys Acta 1796(2):129–139

Layke JC, Lopez PP (2004) Gastric cancer: diagnosis and treatment options. Am Fam Physician 69 (5):1133–1140

Leiserson MD, Vandin F, Wu HT, Dobson JR, Eldridge JV, Thomas JL, Papoutsaki A, Kim Y, Niu B, McLellan M, Lawrence MS (2015) Pan-cancer network analysis identifies combinations of rare somatic mutations across pathways and protein complexes. Nat Genet 47(2):106

Loos RJ, Schadt EE (2012) This I believe: gaining new insights through integrating "old" data. Front Genet 3:137

Lu X, Zhao X, Bai C, Zhao C, Lu G, Xu G (2008) Corrigendum to "LC–MS-based metabonomics analysis" [J Chromatogr B 866 (2008) 64]. J Chromatogr B 1(874):125

Macdonald JS (2006) Gastric cancer--new therapeutic options. N Engl J Med 355(1):76–77

Malley CO, Pidgeon GP (2016) The mTOR pathway in obesity driven gastrointestinal cancers: potential targets and clinical trials. BBA Clin 5:29–40

Mao W, Chen J, Peng TL, Yin XF, Chen LZ, Chen MH (2012) Role of trefoil factor 1 in gastric cancer and relationship between trefoil factor 1 and gastrokine 1. Oncol Rep 28(4):1257–1262
Nakamura R, Saikawa Y, Kubota T, Kumagai A, Kiyota T, Ohashi M, Yoshida M, Otani Y, Kumai K, Kitajima M (2006) Role of the MTT chemosensitivity test in the prognosis of gastric cancer patients after postoperative adjuvant chemotherapy. Anticancer Res 26(2B):1433–1437
Omberg L, Ellrott K, Yuan Y, Kandoth C, Wong C, Kellen MR, Friend SH, Stuart J, Liang H, Margolin AA (2013) Enabling transparent and collaborative computational analysis of 12 tumor types within the cancer genome atlas. Nat Genet 45(10):1121
Paek AL, Liu JC, Loewer A, Forrester WC, Lahav G (2016) Cell-to-cell variation in p53 dynamics leads to fractional killing. Cell 165(3):631–642
Pradhan MP, Prasad NK, Palakal MJ (2012) A systems biology approach to the global analysis of transcription factors in colorectal cancer. BMC Cancer 12(1):331
Pratt D, Chen J, Welker D, Rivas R, Pillich R, Rynkov V, Ono K, Miello C, Hicks L, Szalma S, Stojmirovic A (2015) NDEx, the network data exchange. Cell Syst 1(4):302–305
Rajdev L (2010) Treatment options for surgically resectable gastric cancer. Curr Treat Options Oncol 11(1–2):14–23
Rüschoff J, Hanna W, Bilous M, Hofmann M, Osamura RY, Penault-Llorca F, Van De Vijver M, Viale G (2012) HER2 testing in gastric cancer: a practical approach. Mod Pathol 25(5):637
Safari-Alighiarloo N, Taghizadeh M, Rezaei-Tavirani M, Goliaei B, Peyvandi AA (2014) Protein-protein interaction networks (PPI) and complex diseases. Gastroenterol Hepatol 7(1):17
Saito T, Kondo C, Shitara K, Ito Y, Saito N, Ikehara Y, Yatabe Y, Yamamichi K, Tanaka H, Nakanishi H (2015) Comparison of intratumoral heterogeneity of HER2 expression between primary tumor and multiple organ metastases in gastric cancer: clinicopathological study of three autopsy cases and one resected case. Pathol Int 65(6):309–317
Schoeberl B, Pace EA, Fitzgerald JB, Harms BD, Xu L, Nie L, Linggi B, Kalra A, Paragas V, Bukhalid R, Grantcharova V (2009) Therapeutically targeting ErbB3: a key node in ligand-induced activation of the ErbB receptor–PI3K axis. Sci Signal 2(77):ra31
Serrano L (2007) Synthetic biology: promises and challenges. Mol Syst Biol 3(1):158
Sipponen P, Ranta P, Helske T, Kääriäinen I, Mäki T, Linnala A, Suovaniemi O, Alanko A, Härkönen M (2002) Atrophic gastritis serum levels of amidated gastrin-17 and pepsinogen I in atrophic gastritis: an observational case-control study. Scand J Gastroenterol 37(7):785–791
Song H, Peng JS, Dong-Sheng Y, Yang ZL, Liu HL, Zeng YK, Shi XP, Lu BY (2012) Serum metabolic profiling of human gastric cancer based on gas chromatography/mass spectrometry. Braz J Med Biol Res 45(1):78–85
Starzynska T, Bromley M, Ghosh A, Stern PL (1992) Prognostic significance of p53 overexpression in gastric and colorectal carcinoma. Br J Cancer 66(3):558
Tashiro A, Sano M, Kinameri K, Fujita K, Takeuchi Y (2006) Comparing mass screening techniques for gastric cancer in Japan. World J Gastroenterol 12(30):4873
Uhlén M, Fagerberg L, Hallström BM, Lindskog C, Oksvold P, Mardinoglu A, Sivertsson Å, Kampf C, Sjöstedt E, Asplund A, Olsson I (2015) Tissue-based map of the human proteome. Science 347(6220):1260419
Wang X, Yan SK, Dai WX, Liu XR, Zhang WD, Wang JJ (2010) A metabonomic approach to chemosensitivity prediction of cisplatin plus 5-fluorouracil in a human xenograft model of gastric cancer. Int J Cancer 127(12):2841–2850
Wu L, Qu X (2015) Cancer biomarker detection: recent achievements and challenges. Chem Soc Rev 44(10):2963–2997
Wu H, Xue R, Tang Z, Deng C, Liu T, Zeng H et al (2010) Metabolomic investigation of gastric cancer tissue using gas chromatography/mass spectrometry. Anal Bioanal Chem 396 (4):1385–1395
Yan W, Wang S, Sun Z, Lin Y, Sun S, Chen J, Chen W (2014) Identification of microRNAs as potential biomarker for gastric cancer by system biological analysis. Biomed Res Int 2014:9
Yilmaz M, Christofori G (2010) Mechanisms of motility in metastasizing cells. Mol Cancer Res 8 (5):629–642

The manufacturer's authorised representative in the EU is Springer Nature Customer Service Centre GmbH, Europaplatz 3, 69115 Heidelberg, Germany. If you have any concerns regarding our products, please contact ProductSafety@springernature.com

Printed and bound by CPI Group (UK) Ltd, Croydon, CR0 4YY

24/07/2026

02175947-0001